SECOND EDITION

Basic GIS
Coordinates

SECOND EDITION

Basic GIS
Coordinates

Jan Van Sickle

 **CRC Press**
Taylor & Francis Group
Boca Raton   London   New York

CRC Press is an imprint of the
Taylor & Francis Group, an **informa** business

CRC Press
Taylor & Francis Group
6000 Broken Sound Parkway NW, Suite 300
Boca Raton, FL 33487-2742

Library of Congress Cataloging-in-Publication Data

Van Sickle, Jan
    Basic GIS coordinates / Jan Van Sickle. -- 2nd ed.
      p. cm.
    Includes bibliographical references and index.
    ISBN 978-1-4200-9231-8 (hard back : alk. paper)
    1. Grids (Cartography)--Data processing. 2. Geographic information systems. I. Title.

GA116.V36 2010
910.285--dc22

                                               2010000030

Visit the Taylor & Francis Web site at
http://www.taylorandfrancis.com

and the CRC Press Web site at
http://www.crcpress.com

Contents

Preface

Coordinates? Press a few buttons on a computer and they are automatically imported, exported, rotated, translated, collated, annotated, and served up in any format you choose with no trouble at all. There really is nothing to it. Why have a book about coordinates?

It's a good question, really. Computers are astounding in their ability to make the mathematics behind coordinate manipulation transparent to the user. This book is not much about that sort of mathematics. But it is about coordinates and coordinate systems. It is about understanding how these systems work, and how they sometimes don't work. It is about how points that should be in New Jersey end up in the middle of the Atlantic Ocean even if the computer has done nothing wrong. And that is, I suppose, the answer to the question from my point of view. Computers are currently very good at repetition and very bad at interpretation. People are usually not so good at repetition. We tend to get bored. But we can be very good indeed at interpretation—if we have the information in our heads to understand what we are interpreting. This book is about providing some of that sort of information on the subject of coordinates and coordinate systems.

Coordinates are critical to GIS (geographic information systems), cartography, and surveying. They are their foundation. Points, lines, and polygons form the geometry. Lines define the polygons, points define the lines, and coordinates define the points. So at bottom it is the coordinates that tie the real world to its electronic image in the computer.

There are more than 1,000 horizontal geodetic datums and over 3,000 Cartesian coordinate systems, sanctioned by governments around the world and currently in use, to describe our planet electronically and on paper.

Author

Jan Van Sickle has more than 40 years of experience in GIS (geographic information systems), GPS (global positioning systems), surveying, and mapping. He created and led the GIS department at Qwest Communications for the company's 25,000-mile worldwide fiber-optic network.

He also led the team that built the GIS for natural gas gathering in the Barnett Shale. He began working with GPS in the early 1980s when he supervised control work using the Macrometer, the first commercial GPS receiver. He assisted the supervision of the first GPS control survey of the Grand Canyon. He led the team that collected, processed, and reported GPS ground control positions for more than 120 cities around the world for the orthorectification of satellite imagery now utilized in a global Web utility.

Van Sickle has led nationwide seminars based on his three books, *GPS for Land Surveyors*, *Basic GIS Coordinates*, and *Surveying Solved Problems*. The latter book is serialized in the magazine *POB*. He has been a featured speaker at many conferences, including Geospatial Information & Technology Association (GITA) conferences and the Institute of Navigation (ION) Annual Meeting. He is a senior lecturer at Penn State University. He was formerly on the board of RM-ASPRS (Rocky Mountain Chapter of the American Society for Photogrammetry & Remote Sensing) and he was the vice-chairman of GIS in the Rockies.

Jan earned his Ph.D. in GIS engineering from the University of Colorado. He is a licensed professional land surveyor in Colorado, California, Oregon, Texas, and North Dakota.

1

Foundation of a Coordinate System

Uncertainty

Coordinates are slippery devils. A stake driven into the ground holds a clear position, but it is awfully hard for its coordinates to be so certain, even if the figures are precise. For example, a latitude of 40° 25′ 33.504″ N with a longitude of 108° 45′ 55.378″ W appears to be an accurate, unique coordinate. Actually, it could correctly apply to more than one place. An elevation, or height, of 2658.2 m seems unambiguous too, but it isn't.

In fact, this latitude, longitude, and height once pinpointed a control point known as Youghall, but not anymore. Oh, Youghall still exists. It is a bronze disk cemented into a drill hole in an outcropping of bedrock on Tanks Peak in the Colorado Rocky Mountains. It's not going anywhere, but its coordinates have not been nearly as stable as the monument. In 1937, the U.S. Coast and Geodetic Survey set Youghall at latitude 40° 25′ 33.504″ N and longitude 108° 45′ 55.378″ W. You might think that was that, but in November 1997, Youghall suddenly got a new coordinate, 40° 25′ 33.39258″ N and 108° 45′ 57.78374″ W. That's more than 56 m (185 ft) west and 3 m (11 ft) south of where it started. But Youghall hadn't actually moved at all. Its elevation changed too. It was 2658.2 m in 1937. It is 2659.6 m today. It rose 4½ ft.

Of course, it did no such thing; the station is right where it has always been. The Earth shifted underneath it, well, nearly. It was the *datum* that changed. The 1937 latitude and longitude for Youghall was based on the North American Datum 1927 (NAD27). Seventy years later, in 1997, the basis of the coordinate of Youghall became the North American Datum 1983 (NAD83).

Datums to the Rescue

Coordinates without a specified datum are vague. It means that questions like "Height above what?", "Where is the origin?" and "On what surface do they lie?" go unanswered. When that happens, coordinates are of no use

1

really. An origin, a starting place, is a necessity for them to be meaningful. Not only must they have an origin, they must also be on a clearly defined surface. These foundations constitute the datum.

Without a datum, coordinates are like checkers without a checkerboard: You can arrange them, analyze them, move them around, but absent the framework, you never really know what you've got. In fact, datums—very much like checkerboards—have been in use for a long time. They are generally called Cartesian systems.

René Descartes

Cartesian systems get their name from René Descartes, a mathematician and philosopher. In the world of the seventeenth century, he was also known by the Latin name Renatus Cartesius, which might explain why we have a whole category of coordinates known as Cartesian coordinates. Descartes did not really invent the things, despite a story that he watched a fly walk on his ceiling and then tracked the meandering path with this system of coordinates. Long before, around 250 BC or so, the Greek scholar Eratosthenes used a checkerboard-like grid to locate positions on the Earth, and even he was not the first: Dicaearchus had come up with the same basic idea before him. Nevertheless, Descartes was probably the first to use graphs to plot and analyze mathematical functions. He set up the rules we use now for his particular version of a coordinate system in two dimensions defined on a flat plane by two axes.

Cartesian Coordinates

Cartesian coordinates are expressed in ordered pairs. Each element of the coordinate pair is the distance measured across a flat plane from the point. The distance is measured along the line parallel with one axis that extends to the other axis. If the measurement is parallel with the x-axis, it is called the x-coordinate, and if the measurement is parallel with the y-axis, it is called the y-coordinate.

Figure 1.1 shows two axes perpendicular to each other labeled x and y. This labeling is a custom established by Descartes. His idea was to symbolize unknown quantities with letters from the end of the alphabet—x, y, z. This leaves letters from the beginning available for known values. Coordinates became so often used to solve for unknowns, the principle was established that Cartesian axes have the labels x and y. The fancy names for the axes are *abscissa*, for x, and *ordinate*, for y. Surveyors, cartographers, and mappers call them north and east, but back to the story.

These axes need not be perpendicular to each other. They could intersect at any angle, though they would obviously be of no use if they were parallel. Because so much convenience would be lost using anything other than a right angle, it has become the convention. Another convention is the idea

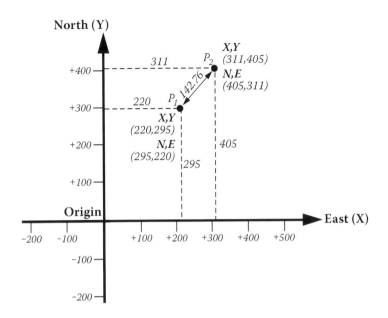

FIGURE 1.1
The Cartesian coordinate system.

that the units along the x-axis are identical with the units along the y-axis, even though there is no theoretical requirement that this be so. Finally, on the x-axis, any point to the west, i.e., to the left, of the origin is negative, and any point to the east, i.e., to the right, is positive. Similarly, on the y-axis, any point north of the origin is positive, and any point south is negative. If these principles are held, then the rules of Euclidean geometry are true, and the off-the-shelf CAD (computer-aided design) and GIS (geographic information system) software on your PC have no trouble at all working with these coordinates; a most practical benefit.

For example, the distance between these points can be calculated using the coordinate geometry you learned in high school. The x- and y-coordinates for the points in the illustration are: the origin point P_1 (220, 295), and point P_2 (311, 405), where:

$X_1 = 220$
$Y_1 = 295$

and

$X_2 = 311$
$Y_2 = 405$

$\text{Distance} = \sqrt{[(X_1 - X_2)^2 + (Y_1 - Y_2)^2]}$

$\text{Distance} = \sqrt{[(220 - 311)^2 + (295 - 405)^2]}$

$$\text{Distance} = \sqrt{[(-91)^2 + (-110)^2]}$$

$$\text{Distance} = \sqrt{(8,281 + 12,100)}$$

$$\text{Distance} = \sqrt{20,381}$$

$$\text{Distance} = 142.76$$

The system works, and it is convenient. But unless it has an attachment to something a bit more real than these unitless numbers, it is not very helpful, which brings up an important point about datums.

Attachment to the Real World

The beauty of datums is that they are errorless, at least in the abstract. On a datum, every point has a unique and accurate coordinate. There is no distortion. There is no ambiguity. For example, the position of any point on the datum can be stated exactly, and it can be accurately transformed into coordinates on another datum with no discrepancy whatsoever. All of these wonderful things are possible only as long as a datum has no connection to anything in the physical world. In that case, it is perfectly accurate, and perfectly useless.

But suppose you wished to assign coordinates to objects on the floor of a very real rectangular room. A Cartesian coordinate system could work, if it is fixed to the room with a well-defined orientation. For example, you could put the origin at the southwest corner, stipulate that the walls of the room are oriented in cardinal directions, and use the floor as the reference plane.

With this datum you not only have the advantage that all of the coordinates are positive, but you can define the location of any object on the floor of the room. The coordinate pairs would consist of two distances, the distance east and the distance north from the origin in the corner. As long as everything stays on the floor, you are in business. In this case, there is no error in the datum, of course, but there are inevitably errors in the coordinates. These errors are due to the less-than-perfect flatness of the floor, the impossibility of perfect measurement from the origin to any object, the ambiguity of finding the precise center of any of the objects assigned coordinates, etc. In short, as soon as you bring in the real world, things get messy.

Cartesian Coordinates and the Earth

Cartesian coordinates then are rectangular, or orthogonal, defined by perpendicular axes from an origin along a specifically oriented reference surface. These elements can define a datum, or framework, for meaningful coordinates.

As a matter of fact, two-dimensional Cartesian coordinates are an important element in the vast majority of coordinate systems; the State Plane Coordinate System (SPCS) in the United States, the Universal Transverse Mercator (UTM) coordinate system, and most others. The datums for these coordinate systems are well established. But there are also local Cartesian coordinate systems whose origins are often entirely arbitrary. For example, if surveying, mapping, or other work is done for the construction of a new building, there may be no reason for the coordinates used to have any fixed relation to any other coordinate systems. In that case, a local datum may be chosen for the specific project, with north and east fairly well defined and with the origin moved far to the west and south of the project to ensure that there will be no negative coordinates. Such an arrangement is good for local work, but it does preclude any easy combination of such small independent systems. Large-scale Cartesian datums, on the other hand, are designed to include positions across significant portions of the Earth's surface into one system. Of course, these are also designed to represent our decidedly round planet on the flat Cartesian plane, no easy task.

But how would a flat Cartesian datum with two axes represent the Earth? There is obviously distortion inherent in the idea. If the planet were flat, it would do nicely, and across small areas, that very approximation—a flat Earth—works reasonably well. That means that even though the inevitable warping involved in representing the Earth on a flat plane cannot be eliminated, it can be kept within well-defined limits as long as the region covered is small and precisely defined. If the area covered becomes too large, distortion does defeat it (Figure 1.2). So the question is, "Why go to all the trouble to work with plane coordinates?" Well, here is a short example to illustrate the utility of this approach.

It is certainly possible to calculate the distance from station Youghall to station Karns using latitude and longitude, also known as *geographic coordinates*, but it is easier for your computer, and for you, to use Cartesian coordinates. Here are the geographic coordinates for these two stations: Youghall at latitude 40° 25' 33.39258" N and longitude 108° 45' 57.78374" W and Karns at latitude 40° 26' 06.36758" N and longitude 108° 45' 57.56925" W in the North American Datum 1983 (NAD83). Here are the same two stations' positions expressed in Cartesian coordinates.

Youghall
$$\text{Northing} = Y_1 = 1{,}414{,}754.47$$
$$\text{Easting} = X_1 = 2{,}090{,}924.62$$

Karns
$$\text{Northing} = Y_2 = 1{,}418{,}088.47$$
$$\text{Easting} = X_2 = 2{,}091{,}064.07$$

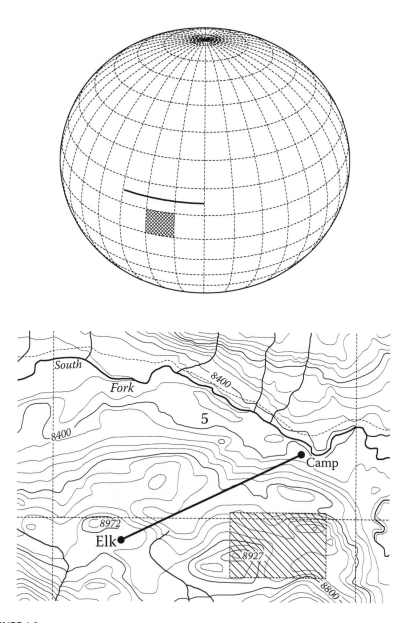

FIGURE 1.2
Distortion of flat systems increases over long lines and large areas.

The Cartesian system used here makes use of state plane coordinates (SPCs) in Colorado's North Zone, and the units are *survey feet* (more about those in Chapter 4). The important point is this: These coordinates are based on a simple two-axis Cartesian system operating across a flat reference plane.

As shown previously, the distance between these points using the plane coordinates is easy to calculate.

$$\text{Distance} = \sqrt{[(X_1 - X_2)^2 + (Y_1 - Y_2)^2]}$$

$$\text{Distance} = \sqrt{[(2,090,924.62 - 2,091,064.07)^2 + (1,414,754.47 - 1,418,088.47)2]}$$

$$\text{Distance} = \sqrt{[(-139.45)^2 + (-3,334.00)^2]}$$

$$\text{Distance} = \sqrt{(19,445.3025 + 11,115,556.0000)}$$

$$\text{Distance} = \sqrt{11,135,001.30}$$

$$\text{Distance} = 3,336.91 \text{ ft}$$

The distance calculated using the plane coordinates is 3336.91 ft. The distance between these points calculated from their latitudes and longitudes is slightly different: 3337.05 ft. Both of these distances are the result of *inverses*, which means that they were calculated between two positions from their coordinates. Comparing the results between the methods shows a difference of about 0.14 ft, a bit more than 0.1 ft. In other words, the spacing between stations would need to grow more than seven times, to about 4½ miles, before the difference would reach 1 ft. So part of the answer to the question, "Why go to all the trouble to work with plane coordinates?" is this: They are easy to use, and the distortion across small areas is not severe.

This rather straightforward idea is behind a good deal of the conversion work done with coordinates. Geodetic coordinates are useful but somewhat cumbersome, at least for conventional trigonometry. Cartesian coordinates on a flat plane are simple to manipulate but inevitably include distortion. Moving from one to the other, it is possible to gain the best of both. The question is, "How do you project coordinates from the nearly spherical surface of the Earth to a flat plane?" Well, first you need a good model of the Earth.

The Shape of the Earth

People have been proposing theories about the shape and size of the planet for a couple of thousand years. Despite the fact that local topography is obvious to an observer standing on the Earth, efforts to grasp the more general nature of the planet's shape and size have been occupying scientists for at least 2,300 years. There have, of course, been long intervening periods of unmitigated nonsense on the subject. Ever since 200 BC, when Eratosthenes

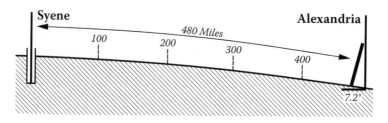

FIGURE 1.3
Eratosthenes' data.

almost calculated the planet's circumference correctly (Figure 1.3), geodesy has been getting ever closer to expressing the actual shape of the Earth in numerical terms.

At noon, the reflection of the midsummer sun was there in the water of a deep well at Syene. The sun was directly overhead. On the same day, the noon shadow cast by a pillar at Alexandria was measured. It showed that the sunbeam strikes the earth at an angle of 7.2° off the vertical. Therefore, the angle between Alexandria and Syene must be 7.2°—1/50th of a 360° circle. Syene is 480 miles south of Alexandria, and a great circle must therefore be 50 times 480 miles in length, or 24,000 miles. In fact, the circumference of the earth is around 24,900 miles.

A real breakthrough came in 1687 when Newton suggested that the Earth shape was ellipsoidal in the first edition of his *Principia*. The idea was not entirely without precedent. Years earlier, astronomer J. Richter found that the closer he got to the equator, the more he had to shorten the pendulum on his 1-sec clock. It swung more slowly in French Guiana than it did in Paris. When Newton heard about it, he speculated that the force of gravity was less in South America than in France. He explained the weaker gravity by the proposition that, when it comes to the Earth, there is simply more of it around the equator. He wrote, "The Earth is higher under the equator than at the poles, and that by an excess of about 17 miles" (*Philosophiae naturalis principia mathematica*, Book III, Proposition XX). He was pretty close to right; the actual distance is only about 4 miles less than he thought.

Some supported Newton's idea that the planet bulged around the equator and flattened at the poles, but others disagreed, including the director of the Paris Observatory, Jean Dominique Cassini. Even though he had seen the flattening of the poles of Jupiter in 1666, neither he nor his son Jacques were prepared to accept the same idea when it came to the Earth. And it appeared they had some evidence on their side.

For geometric verification of the Earth model, scientists had employed arc measurements since the early 1500s. First, they would establish the latitude of their beginning and ending points astronomically. Next, they would measure north along a meridian and find the length of one degree of latitude along that longitudinal line. Early attempts assumed a spherical Earth, and the results were used to estimate its radius by simple multiplication. In fact, one of the most accurate of the measurements of this type, begun in 1669 by the French abbé J. Picard, was actually used by Newton in formulating his own law of gravitation. However, Cassini noted that close analysis of Picard's arc measurement, and measurements of others, seemed to show that the length along a meridian through one degree of latitude actually *decreased* as it proceeded northward. If that was true, then the Earth was elongated at the poles, not flattened.

The argument was not resolved until the famous Swedish physicist Anders Celsius, on a visit to Paris, suggested two expeditions. One group, led by Moreau de Maupertuis, went to measure a meridian arc along the Tornio River near the Arctic Circle, at 66° 20′ north latitude, in Lapland. Another expedition went to what is now Ecuador to measure a similar arc near the equator, 01° 31′ south latitude. The Tornio expedition reported that 1° along the meridian in Lapland was 57,437.9 toises, about 69.6 miles. A *toise* is approximately 6.4 ft. A degree along a meridian near Paris had been measured as 57,060 toises, about 69.1 miles. This shortening of the length of the arc was taken as proof that the Earth is flattened near the poles. Even though the measurements were wrong, the conclusion was correct. Maupertuis published a book on the work in 1738, the king of France gave Celsius a yearly pension of 1,000 livres, and Newton's conjecture was proved right. I wonder which of them was the most pleased.

Since then, there have been numerous meridian measurements all over the world, not to mention satellite observations, and it is now settled that the Earth most nearly resembles an oblate spheroid. An oblate spheroid is an ellipsoid of revolution. In other words, it is the solid generated when an ellipse is rotated around its shorter axis and then flattened at its poles. The flattening is only about 1 part in 300. Still, the ellipsoidal model, bulging at the equator and flattened at the poles, is the best representation of the general shape of the Earth. If such a model of the Earth were built with an equatorial diameter of 25 ft, the polar diameter would be about 24 ft 11 in., almost indistinguishable from a sphere.

It is on this somewhat ellipsoidal Earth model that latitude and longitude have been used for centuries. The idea of a nearly spherical grid of imaginary intersecting lines has helped people navigate around the planet for more than 1,000 years and is showing no signs of wearing down. It is still a convenient and accurate way of defining positions.

Latitude and Longitude

Latitude and longitude are coordinates that represent a position on the surface of the Earth with angles instead of distances. Usually the angles are measured in degrees, but *grads* and *radians* are also used. Depending on the precision required, the degrees (360° comprising a full circle) can be subdivided into 60 min of arc, and each minute of arc can be further subdivided into 60 sec of arc. In other words, there are 3600 sec in a degree. Seconds can be subsequently divided into decimals of seconds. Typically, the arc is dropped from their names, since it is usually obvious that the minutes and seconds are in space rather than time. In any case, these subdivisions are symbolized by

° for degrees

′ for minutes

″ for seconds

The system is called *sexagesimal*. A radian is the angle subtended by an arc equal to the radius of a circle. A full circle is 2π radians, and a single radian is 57° 17′ 44.8″.

In the European *centesimal* system, a full circle is divided into 400 grads. These units are also known as *grades* and *gons*. As in the sexagesimal system, a radian is the angle subtended by an arc equal to the radius of a circle. A full circle is 2π radians, and a single radian is 57° 17′ 44.8″.

Lines of latitude and longitude always cross each other at right angles, as do the lines of a Cartesian grid, but latitude and longitude exist on a curved rather than a flat surface. There is imagined to be an infinite number of these lines on the ellipsoidal model of the Earth. In other words, each and every place has a line of latitude and a line of longitude passing through it, and it takes both of them to fully define a place. If the distance from the surface of the ellipsoid is then added to the latitude and longitude, you have one type of three-dimensional coordinate. This distance component is sometimes measured as the elevation above the ellipsoid, also known as the ellipsoidal height, and sometimes it is measured all the way from the center of the ellipsoid (see Chapter 3). For the moment, the discussion will be confined to positions on an ellipsoidal model of the Earth with a smooth surface. In this case, the height component can be set aside with the assertion that all positions are on the surface of the model.

In mapping, latitude is usually represented by the lowercase Greek letter phi (φ). Longitude is usually represented by the lowercase Greek letter lambda (λ). In both cases, the angles originate at a plane that is imagined to intersect the ellipsoid. In both latitude and longitude, the planes of origination are intended to include the center of the Earth. Angles of latitude most often originate at the plane of the equator, and angles of longitude originate at the plane through an arbitrarily chosen place, currently Greenwich, England. Latitude is an angular measurement of the extent to which a particular point lies north or south of the equatorial plane measured in degrees, minutes, and seconds (and, usually, decimals of a second). Longitude is also an angle measured in degrees, minutes, seconds, and decimals of a second east and west of the plane through the chosen prime, or zero, position.

Between the Lines

On the Earth, any two lines of longitude—for example, west longitude 89° 00′ 00″ and west longitude 90° 00′ 00″—are farthest from each other at the equator, but as they proceed north and south to the poles, they become closer. In other words, they *converge*. It is interesting to note that the length of a degree of longitude and the length of a degree of latitude are just about the same in the vicinity of the equator. They are both about 60 nautical miles: approximately 111 km (69 miles). But if you imagine going north or south along a line of longitude toward either the North or the South Pole, a degree of longitude becomes progressively shorter. At 2/3 of the distance from the equator to the pole, i.e., at 60° north and south latitudes, a degree of longitude shrinks to about 55.5 km (34.5 miles)—half the length it had at the equator. And as one proceeds northward or southward, a degree of longitude continues to shrink until it fades away to nothing, as shown in Figure 1.4.

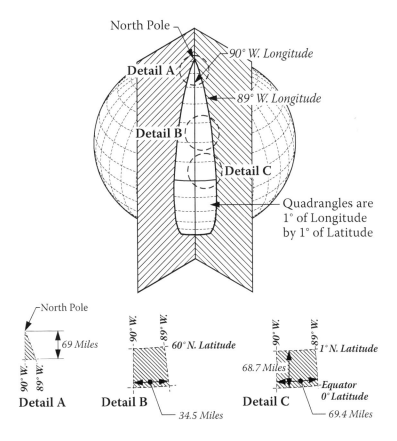

FIGURE 1.4
Distances across 1°.

On the other hand, lines of latitude do not converge on the Earth; they are always parallel to one another and the equator. In fact, as one approaches the poles, where a degree of longitude becomes small, a degree of latitude actually grows slightly. This small increase is due to the flattening near the poles mentioned in the previous discussion of the oblate shape of the planet. The increase in the size of a degree of latitude would not happen if the Earth were a sphere; in that case the length of a degree of latitude would consistently be approximately 110.6 km (68.7 miles), as it is near the equator. However, since the Earth is an oblate spheroid, as Newton predicted, a degree of latitude actually gets a bit longer at the poles. It grows to about 111.7 km (69.4 miles) in that region, which is what all those scientists were trying to measure back in the eighteenth century.

Longitude

Longitude is an angle between two planes. It is a *dihedral angle*. In other words, it is an angle measured at the intersection of two planes that are themselves perpendicular to the equator. In the case of longitude, the first plane passes through the point of interest, the place whose longitude you wish to know, and the second plane passes through an arbitrarily chosen point representing zero longitude. Today, that place is Greenwich, England. The measurement of angles of longitude is imagined to take place where the two planes meet, and that place is the line known as the polar axis. As it happens, that line is also the axis of rotation of the aforementioned ellipsoidal model of the Earth. And where they intersect that ellipsoidal model, they create an elliptical line on its surface. This elliptical line is then divided into two *meridians* at the polar axis. One half becomes a meridian of east longitude, which is labeled E or given a positive (+) value, and the other half becomes a meridian of west longitude, which is labeled W or given a negative (–) value, as shown in Figure 1.5.

The zero meridian through Greenwich is called the *prime meridian*. From there, meridians range from +0° to +180° E longitude and from –0° to –180° W longitude. Taken together, these meridians cover the entire 360° around the Earth. This arrangement was one of the decisions made by consensus of 25 nations in 1884.

The location of the prime meridian is arbitrary. The idea that it passes through the principal transit instrument, the main telescope, at the Royal Observatory at Greenwich, England, was formally established at the International Meridian Conference in Washington, D.C., where it was decided that there would be a single zero meridian rather than the many used before. There were several other decisions made at the meeting, and among them was the agreement that all longitude would be calculated both east and west from the prime meridian up to 180°, with east longitude being positive and west longitude being negative.

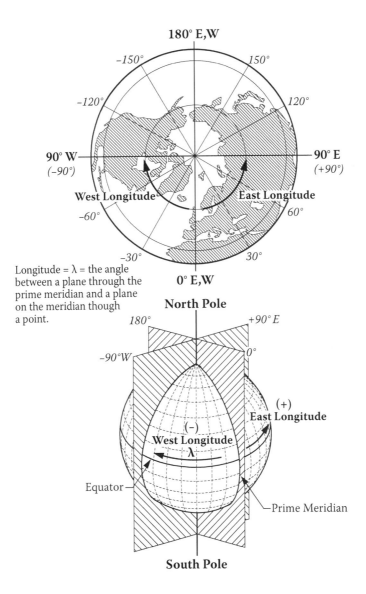

FIGURE 1.5
Longitude.

The 180° meridian is a unique longitude; like the prime meridian, it divides the Eastern Hemisphere from the Western Hemisphere, but it also represents the International Date Line. The calendars west of the line are one day ahead of those east of the line. This division could theoretically occur anywhere on the globe, but it is convenient for it to be 180° from Greenwich in a part of the world mostly covered by ocean. Even though the line does not actually follow the meridian exactly, it avoids dividing populated areas; it illustrates the relationship between longitude and time. Since there are 360° of longitude and 24 hours in a day, it follows that the Earth must rotate at a rate of 15° per hour. This is an idea that is inseparable from the determination of longitude.

Latitude

Two angles are sufficient to specify any location on the reference ellipsoid representing the Earth. Latitude is an angle between a plane and a line through a point.

Imagine a flat plane intersecting an ellipsoidal model of the Earth. Depending on exactly how it is done, the resulting intersection would be either a circle or an ellipse, but if the plane is coincident with or parallel to the equator, the result is always a *parallel of latitude*. The equator is a unique parallel of latitude that also contains the center of the ellipsoid, as shown in Figure 1.6.

The equator is 0° latitude, and the North and South Poles +90° north and −90° south latitude, respectively. In other words, values for latitude range from a minimum of 0° to a maximum of 90°. The latitudes north of the equator are positive and those to the south are negative.

Lines of latitude are called *parallels* because they are always parallel to each other as they proceed around the globe. They do not converge as meridians do or cross each other.

Categories of Latitude and Longitude

When positions given in latitude and longitude are called geographic coordinates, this general term really includes several types. For example, there are geocentric and geodetic versions of latitude and longitude.

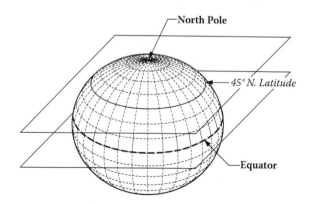

FIGURE 1.6
Parallels of latitude.

The geodetic longitude of a point is the angle between the plane of the Greenwich meridian and the plane of the meridian that passes through the point of interest, both planes being perpendicular to the equatorial plane. Since they have the same zero meridian and the same axis, geodetic longitude and geocentric longitude are equivalent, but when it comes to latitude that is not the case.

It is the ellipsoidal nature of the model of the Earth that contributes to the difference. For example, there are just a few special circumstances on an ellipsoid where a line from a particular position can be both perpendicular to the ellipsoid's surface and also pass through the center. Lines from the poles and lines from the equatorial plane can do that, but in every other case, a line can either be perpendicular to the surface of the ellipsoid, or it can pass through the center, but it cannot do both. And there you have the basis for the difference between geocentric and geodetic latitude.

Imagine a line from the point of interest on the ellipsoid to the center of the Earth. The angle that line makes with the equatorial plane is the point's geocentric latitude. On the other hand, geodetic latitude is derived from a line that is perpendicular to the ellipsoidal model of the Earth at the point of interest. The angle this line makes with the equatorial plane of that ellipsoid is called geodetic latitude. As you can see, geodetic latitude is always just a bit larger than geocentric latitude except at the poles and the equator, where they are the same. The maximum difference between geodetic and geocentric latitude is about 11′44″ and occurs at about 45°.

When latitude and longitude are mentioned without a particular qualifier, in most cases it is best to presume that the reference is to geodetic latitude and longitude (Figure 1.7).

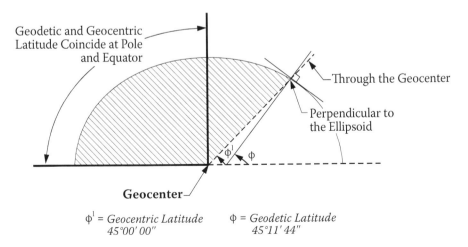

ϕ^1 = *Geocentric Latitude*
45°00′00″

ϕ = *Geodetic Latitude*
45°11′44″

FIGURE 1.7
Geocentric and geodetic latitude.

The Deflection of the Vertical

Down seems like a pretty straightforward idea. A hanging plumb bob certainly points down. Its string follows the direction of gravity. That is one version of the idea. There are others.

Imagine an optical surveying instrument set up over a point. If it is centered precisely with a plumb bob and leveled carefully, the plumb line and the line of the level telescope of the instrument are perpendicular to each other. In other words, the level line, the horizon of the instrument, is perpendicular to gravity. Using an instrument so oriented, it is possible to determine the latitude and longitude of the point. Measuring the altitude of a circumpolar star is one good method of finding the latitude. The measured altitude would be relative to the horizontal level line of the instrument. Latitude found this way is called *astronomic latitude*.

One might expect that this astronomic latitude would be the same as the geocentric latitude of the point, but they are different. The difference is due to the fact that a plumb line coincides with the direction of gravity; it does not point to the center of the Earth, where the line used to derive geocentric latitude originates.

Astronomic latitude also differs from the most widely used version of latitude: the *geodetic latitude*. The line from which geodetic latitude is determined is perpendicular to the surface of the ellipsoidal model of the Earth. That does not match a plumb line either. In other words, there are three different versions of "down," each with its own latitude. For geocentric latitude, down is along a line to the center of the Earth. For geodetic latitude, down is along a line perpendicular to the ellipsoidal model of the Earth. For astronomic latitude, down is along a line in the direction of gravity. More often than not, these are three completely different lines, as seen in Figure 1.8.

Each can be extended upward too, toward the zenith, and there are small angles between them. The angle between the vertical extension of a plumb line and the vertical extension of a line perpendicular to the ellipsoid is called the *deflection of the vertical*. It sounds better than the "difference in down." This deflection of the vertical defines the actual angular difference between the astronomic latitude and longitude of a point and its geodetic latitude and longitude. This applies to both latitude and longitude because, even though the discussion has so far been limited to latitude, the deflection of the vertical usually has both a north-south and an east-west component.

It is interesting to note that that optical instrument, set up so carefully over a point on the Earth, cannot be used to measure geodetic latitude and longitude directly because they are not relative to the actual Earth, but rather a model of it. Gravity does not even come into the ellipsoidal version of down. On the model of the Earth, down is a line perpendicular to the ellipsoidal surface at a particular point. On the real Earth, down is the direction of gravity at the point. They are most often not the same. And since the model is

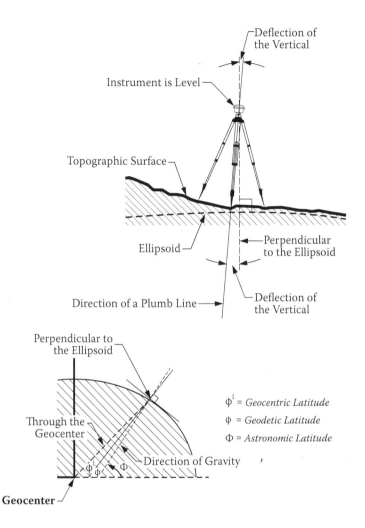

FIGURE 1.8
Geocentric, geodetic, and astronomic latitude.

imaginary, it is quite impossible to actually set up an instrument on the ellipsoid. On the other hand, astronomic observations for the measurement of latitude and longitude by observing stars and planets with instruments on the real Earth have a very long history indeed. And yet the most commonly used coordinates are not astronomic latitudes and longitudes, but geodetic latitudes and longitudes. So conversion from astronomic latitude and longitude to geodetic latitude and longitude has a long history as well. Therefore, until the advent of GPS, geodetic latitudes and longitudes were often values ultimately derived from astronomic observations by postobservation calculation. And in a sense that is still true; the change is that a modern GPS receiver can display the geodetic latitude and longitude of a point to the user immediately because the calculations can be completed with incredible speed. But a fundamental fact remains unchanged: The instruments by which latitudes and longitudes are measured are oriented to gravity, while the ellipsoidal model on which geodetic latitudes and longitudes are determined is not. And that is just as true for the antenna of a GPS receiver, an optical surveying instrument, a camera in an airplane taking aerial photography, or even the GPS satellites themselves.

To illustrate the effect of the deflection of the vertical on latitude and longitude, here are station Youghall's astronomical latitude and longitude labeled with capital phi (Φ) and capital lambda (Λ), respectively, the standard Greek letters commonly used to differentiate them from geodetic latitude and longitude:

$\Phi = 40°\ 25'\ 36.28''\ N$

$\Lambda = 108°\ 46'\ 00.08''\ W$

Now, the deflection of the vertical can be used to convert these coordinates to a geodetic latitude and longitude. Unfortunately, the small angle is not usually conveniently arranged. It would be helpful if the angle between the direction of gravity and the perpendicular to the ellipsoid would follow only one cardinal direction, north-south or east-west. For example, if the angle observed from above Youghall was oriented north or south along the meridian, then it would affect only the latitude (not the longitude) and would be very easy to apply. But that is not the case. The two *normals*, i.e., the perpendicular lines that constitute the deflection of the vertical at a point, are usually neither north-south nor east-west of each other. Looking down on a point, one could imagine that the angle they create between them stands in one of the four quadrants: northeast, southeast, southwest, or northwest. Therefore, to express its true nature, deflection is broken down into two components, one north-south and the other east-west. There are almost always some of both. The north-south component is known by the Greek letter xi (ξ). It is positive (+) to the north and negative (–) to the south. The east-west component is known by the Greek letter eta (η). It is positive (+) to the east and negative (–) to the west.

For example, as seen in Figure 1.9, the components of the deflection of the vertical at Youghall are:

North-south = xi = ξ = +2.89''

East-west = eta = η = +1.75''

In other words, if an observer held a plumb bob directly over the monument at Youghall, the upper end of the string would be 2.89 arc-sec north and 1.75 arc-sec east of the line that is perpendicular to the ellipsoid.

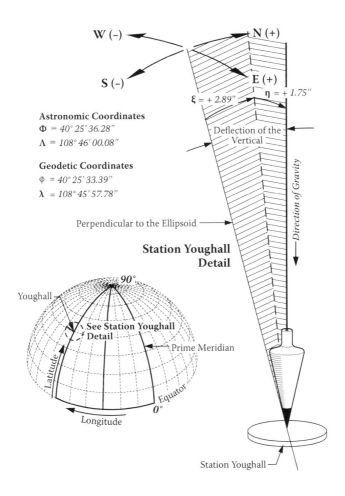

FIGURE 1.9
Two components of the deflection of the vertical at station Youghall.

The geodetic latitude and longitude (φ and λ) can be computed from the astronomic latitude and longitude (Φ and Λ) using the following formulas:

$\varphi = \Phi - \xi$
$\varphi = 40°\ 25'\ 36.28'' - (+2.89'')$
$\varphi = 40°\ 25'\ 33.39''$

$\lambda = \Lambda - \eta/\cos\varphi$
$\lambda = 108°\ 46'\ 00.08'' - (+1.75'')/\cos 40°\ 25'\ 33.39''$
$\lambda = 108°\ 46'\ 00.08'' - (+1.75'')/0.7612447$
$\lambda = 108°\ 46'\ 00.08'' - (+2.30'')$
$\lambda = 108°\ 45'\ 57.78''$

where

  φ = geodetic latitude
  λ = geodetic longitude
  Φ = astronomical latitude
  Λ = astronomical longitude

and the components of the deflection of the vertical are

  North-south = ξ
  East-west = η

Directions

Azimuths

An *azimuth* is one way to define the direction from point to point on the ellipsoidal model of the Earth, on Cartesian datums, and on other systems. On some Cartesian datums, an azimuth is called a *grid* azimuth, referring to the rectangular grid on which a Cartesian system is built. Grid azimuths are defined by a horizontal angle measured clockwise from north.

Azimuths can be measured clockwise from north either through a full 360° or measured +180° in a clockwise direction from north and −180° in a counterclockwise direction from north. Bearings are different.

Bearings

Bearings, another method of describing directions, are always acute angles measured from 0° at either north or south to 90° to either the west or the east. They are measured both clockwise and counterclockwise. They are expressed from 0° to 90° from north in two of the four quadrants, the northeast, 1, and northwest, 4. Bearings are also expressed from 0° to 90° from south in the two remaining quadrants, the southeast, 2, and southwest, 3 as shown in Figure 1.10.

In other words, bearings use four quadrants of 90° each. A bearing of N 45° 15′ 35″ E is an angle measured in a clockwise direction 45° 15′ 35″ from north toward the east. A bearing of N 21° 44′ 52″ W is an angle measured in a counterclockwise direction 21° 44′ 52″ toward west from north. The same ideas work for southwest bearings measured clockwise from south and southeast bearings measured counterclockwise from south.

Directions—azimuths and bearings—are indispensable. They can be derived from coordinates with an inverse calculation. If the coordinates of two points inversed are geodetic, then the azimuth or bearing derived from

VEE 309-8

COLLECT BY: 2 2 JAN 2018

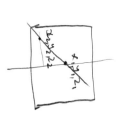

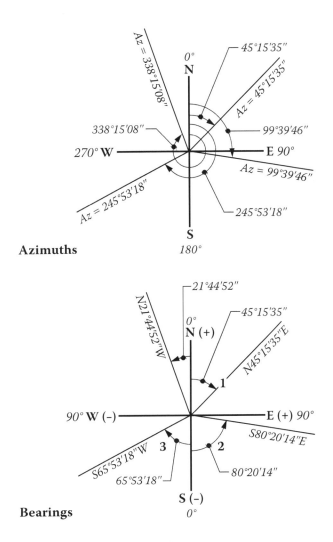

FIGURE 1.10
Azimuths and bearings.

them is also geodetic; if the coordinates are astronomic, then the direction will be astronomic, and so on. If the coordinates from which a direction is calculated are *grid coordinates*, the resulting azimuth will be a grid azimuth, and the resulting bearing will be a grid bearing.

Both bearings and azimuths in a Cartesian system assume the direction to north is always parallel to the *y*-axis, the north-south axis. On a Cartesian datum, there is no consideration for convergence of *meridional*, north-south, lines. One result of the lack of convergence is that the bearing or azimuth at one end of a line is always exactly 180° different from the bearing or azimuth at the other end of the same line. But if the datum is on the ellipsoidal model of the Earth, directions do not quite work that way. For example, consider the difference between an astronomic azimuth and a geodetic azimuth.

Astronomic and Geodetic Directions

If it were possible to point an instrument to the exact position of the north celestial pole, a horizontal angle turned from there to an observed object on the Earth would be the astronomic azimuth to that object from the instrument. But it is rather difficult to measure an astronomic azimuth that way because there is nothing to point to at the celestial North Pole but a lot of sky. Polaris, the North Star, appears to follow an elliptical path around the celestial North Pole, which is the northward prolongation of the Earth's axis. Even so, Polaris and several other celestial bodies, for that matter, serve as good references for the measurement of astronomic azimuths, albeit with a bit of calculation. Still, optical instruments used to measure astronomic azimuths must be oriented to gravity, and it is usual for the azimuths derived from celestial observations to be converted to geodetic azimuths. Geodetic azimuths are the native forms on an ellipsoid. So, as it was with the conversion of astronomic latitudes and longitudes to geodetic coordinates, the deflection of the vertical is applied to convert astronomic azimuths to geodetic azimuths.

For example, the astronomic azimuth between two stations, from Youghall to Karns, is 00° 17′ 06.67″. Given this information, the geodetic latitude of Youghall, 40° 25′ 33.39″ N, and the east-west component of the deflection of the vertical, η = +1.75, it is possible to calculate the geodetic azimuth from Youghall and Karns using the following formula:

$\alpha = A - \eta \tan \varphi$

$\alpha = 00° 17′ 06.67″ - (+1.75″) \tan 40° 25′ 33.39″$

$\alpha = 00° 17′ 06.67″ - (+1.75″) 0.851847724$

$\alpha = 00° 17′ 06.67″ - 1.49$

$\alpha = 00° 17′ 05.18″$

where:

α = geodetic azimuth
A = astronomical azimuth
η = the east-west component of the deflection of the vertical
φ = geodetic latitude

But, as always, there is another way to calculate the difference between an astronomic azimuth and a geodetic azimuth. Here is the formula and a calculation using the data at Youghall:

Φ = 40° 25′ 36.28″ N
Λ = 108° 46′ 00.08″ W

$\varphi = $ 40° 25′ 33.39″ N
$\lambda = $ 108° 45′ 57.78″ W

$\alpha_A - \alpha_G = +(\Lambda - \lambda)\sin \varphi$
$\alpha_A - \alpha_G = +(108° \ 46′ \ 00.08″ - 108° \ 45′ \ 57.78″)\sin 40° \ 25′ \ 33.39″$
$\alpha_A - \alpha_G = +(2.30″)\sin 40° \ 25′ \ 33.39″$
$\alpha_A - \alpha_G = +(2.30″)0.64846$
$\alpha_A - \alpha_G = +1.49″$

where:
   $\alpha_A = $ astronomic azimuth
   $\alpha_G = $ geodetic azimuth
   $\Lambda = $ astronomic longitude
   $\lambda = $ geodetic longitude
   $\varphi = $ geodetic latitude

Even without specific knowledge of the components of the deflection of the vertical, it is possible to calculate the difference between an astronomic azimuth and a geodetic azimuth. The required information is in the coordinates of the point of interest. Knowing the astronomic longitude and the geodetic latitude and longitude of the position is all that is needed. This method of deriving a geodetic azimuth from an astronomic observation makes it easy for surveyors to derive the *LaPlace correction,* which is the name given to the right-hand side of the previous equation.

North

The reference for directions is north, and each category refers to a different north. Geodetic north differs from astronomic north, which differs from grid north, which differs from magnetic north. The differences between the geodetic azimuths and astronomic azimuths are a few seconds of arc at a given point. Variations between these two are small indeed compared with those found between grid azimuths and magnetic azimuths. For example, while there might be a few seconds between astronomic north and geodetic north, there is usually a difference of several degrees between geodetic north and magnetic north.

Magnetic North

Magnetic north is used throughout the world as the basis for magnetic directions in both the Northern and the Southern Hemispheres, but it will not hold still. The position of the magnetic North Pole is somewhere around 79° N latitude, and 106° W longitude, a long way from the geographic North Pole. To make matters even more interesting, the magnetic North Pole is moving at a rate of about 15 miles per year, just a bit faster than it used to. In fact, it has moved more than 600 miles since the early nineteenth century.

The Earth's magnetic field is variable. For example, if the needle of a compass at a particular place points 15° west of geodetic north. There is said to be a west *declination* of 15°. At the same place 20 years later, that declination may have grown to 16° west of geodetic north. This kind of movement is called *secular variation*. It is a change that occurs over long periods and is probably caused by convection in the material at the Earth's core. Declination is one of the two major categories of magnetic variation. The other magnetic variation is called daily or *diurnal* variation.

Daily variation is probably due to the effect of the solar wind on the Earth's magnetic field. As the Earth rotates, a particular place alternately moves toward and away from the constant stream of ionized particles from the Sun. Therefore, it is understandable that the daily variation swings from one side of the mean declination to the other over the course of a day. For example, if the mean declination at a place were 15° west of geodetic north, it might be 14.9° at 8 a.m., 15.0° at 10 a.m., 15.6° at 1 p.m., and again 15.0° at sundown. Such a diurnal variation would be somewhat typical, but in high latitudes it can grow as large as 9°.

Grid North

The position of magnetic north is governed by natural forces, but grid north is entirely artificial. In Cartesian coordinate systems, whether known as *State Plane, Universal Transverse Mercator*, a local assumed system, or any other, the direction to north is established by choosing one meridian of longitude. The meridian that is chosen is usually in the middle of the area, the *zone*, that is covered by the coordinate system. That is why it is frequently known as the *central meridian*. Thereafter, throughout the system, at all points, grid north is along a line parallel to that central meridian. This arrangement purposely ignores the fact that a different meridian passes through each of the points and that all the meridians inevitably converge with one another. Actually, grid north and geodetic north only agree at points along the central meridian; everywhere else in the coordinate system there is an angular difference between the two directions. That angular difference is known as the *convergence*. East of the central meridian grid north is east of geodetic north, and the convergence is positive. West of the central meridian grid north is west of geodetic north, and the convergence is negative. The approximate grid azimuth of a line is its geodetic azimuth minus the convergence. Therefore, it follows that, east of the central meridian, the grid azimuth of a line is smaller than its geodetic azimuth. West of the central meridian, the grid azimuth of a line is larger than the geodetic azimuth, as shown in Figure 1.11.

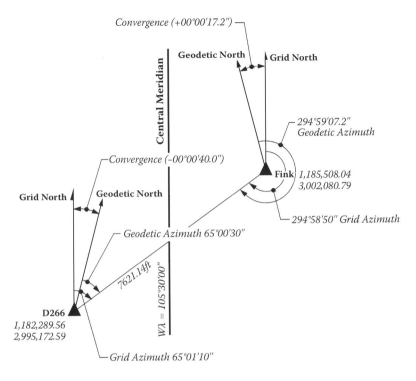

FIGURE 1.11
Approximate grid azimuth = geodetic azimuth – convergence.

Polar Coordinates

There is another way of looking at a direction: It can be one component of a coordinate.

A procedure familiar to surveyors using optical instruments involves the occupation of a station with an established coordinate. A back sighting is taken either on another station with a coordinate on the same datum or on some other reference such as Polaris. With two known positions, the occupied and the sighted, a beginning azimuth or bearing is calculated. Next, a new station is sighted ahead, fore-sighted, on which a new coordinate will be established. The angle is measured from the back sight to the fore sight, fixing the azimuth or bearing from the occupied station to the new station. A distance is measured to the new station, and this direction and distance together can actually be considered the coordinate of the new station. They constitute what is known as a *polar coordinate*. In surveying, polar coordinates are very often a first step toward calculating coordinates in other systems.

There are coordinates that are all distances: Cartesian coordinates, for example. There are coordinates that are all angles: latitude and longitude, for

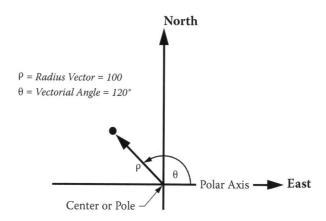

FIGURE 1.12
A polar coordinate (mathematical convention).

example. Then there are coordinates that represent an angle and a distance: polar coordinates, as shown in Figure 1.12.

A polar coordinate defines a position with an angle and distance. As in a Cartesian coordinate system, they are reckoned from an origin, which in this case is also known as the *center* or the *pole*. The angle used to define the direction is measured from the *polar axis*, which is a fixed line pointing to the east, in the configuration used by mathematicians. It is notable that many disciplines, like computer-aided drafting utilities, presume east to be the reference line for directions. Mappers, cartographers, and surveyors tend to use north as the reference for directions in polar coordinates.

In the typical format for recording polar coordinates, the Greek letter rho (ρ) indicates the length of the *radius vector*, i.e., the line from the origin to the point of interest. The angle from the polar axis to the radius vector is represented by the Greek letter theta (θ) and is called the *vectorial angle*, the *central angle*, or the *polar angle*. These values ρ and θ are given in ordered pairs, like Cartesian coordinates. The length of the radius vector is first and the vectorial angle second: for example (100,220°).

There is a significant difference between Cartesian coordinates and polar coordinates. In an established datum using Cartesian coordinates, one and only one ordered pair can represent a particular position. Any change in either the northing or the easting and the coordinate represents a completely different point. However, in the mathematician's polar coordinates, the same position might be represented in many different ways, with many different ordered pairs of ρ and θ standing for the very same point. For example, (87,45°) can just as correctly be written (87,405°), as illustrated in (i) in Figure 1.13. Here the vectorial angle swings through 360° and continues past the pole another 45°. It could also be written as (87,–315), as illustrated in (ii) in Figure 1.13. When θ has a clockwise rotation from the

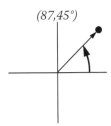

(87,45°)

M a y   a l s o   b e   e x p r e s s e d   a s:

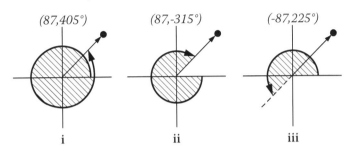

(87,405°)	(87,-315°)	(-87,225°)
i	ii	iii

FIGURE 1.13
Four ways of noting one position.

polar axis in this arrangement, it is negative. Another possibility is a positive or counterclockwise rotation from the polar axis to a point 180° from the origin and the radius vector is negative (−87,225°). The negative radius vector indicates that it proceeds out from the origin in the opposite direction from the end of the vectorial angle, as shown in (iii) in Figure 1.13.

In other words, there are several ways to represent the same point in polar coordinates. This is not the case for rectangular coordinates, nor is it the case for the polar coordinate system as used in surveying, mapping, and cartography. In mapping and cartography, directions are consistently measured from north and the polar axis points north, as shown in Figure 1.14.

In the mathematical arrangement of polar coordinates, a counterclockwise vectorial angle θ is positive and a clockwise rotation is negative. In the surveying, mapping, and cartography arrangement of polar coordinates, the opposite is true. A counterclockwise rotation is negative and a clockwise rotation is positive. The angle may be measured in degrees, radians, or grads, but if it is clockwise, it is positive.

In the mathematical arrangement, the radius vector can be positive or negative. If the point P lies in the same direction as the vectorial angle, it is considered positive. If the point P lies in the opposite direction back through the origin, the radius vector is considered negative. In the surveying, mapping, and cartography arrangement of polar coordinates, the radius vector always points out from the origin and is always positive.

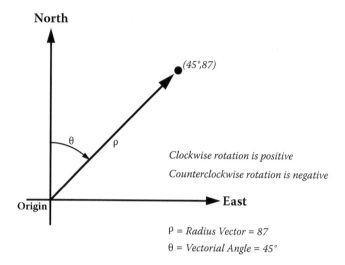

FIGURE 1.14
A polar coordinate (mapping and surveying convention).

Summary

Positions in three-dimensional space can be expressed in both Cartesian coordinates and polar coordinates by the addition of a third axis, the z-axis. The z-axis is perpendicular to the plane described by the x-axis and the y-axis. The addition of a third distance in the Cartesian system, or the addition of a second angle in the polar coordinate system, completes the three-dimensional coordinates of a point.

The letters φ' and λ' represent the two angles in Figure 1.15. If the origin of the axes is placed at the center of an oblate ellipsoid of revolution, the result is a substantially correct model of the Earth from which three-dimensional polar coordinates can be derived. Of course, there is a third element to the polar coordinate here represented by rho, ρ. However, if every position coordinated in the system is always understood to be on the surface of the Earth, or a model of the Earth, this radius vector can be dropped from the coordinate without creating ambiguity. And that is conventionally done, so one is left with the idea that each latitude and longitude comprises a three-dimensional polar coordinate with only two angular parts. In Figure 1.15, they are geocentric latitude and longitude.

If the three-dimensional polar coordinates represent points on the actual surface of the Earth, the irregularity of the planet's surface presents problems. If they represent points on an ellipsoidal model of the Earth, they are on a regular surface, but that surface does not stand at a constant radial distance from the center of the ellipsoid. There is also a difficulty regarding the origin. If the intersection of the axes is at the geocenter, one can derive geocentric latitude and longitude directly. However, the vector perpendicular to

Polar Coordinate

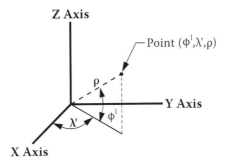

Cartesian Coordinate

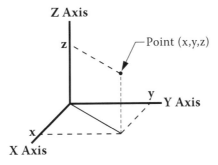

FIGURE 1.15
Three-dimensional polar and Cartesian coordinates.

the surface of the ellipsoid that represents the element of a geodetic latitude and longitude of a point does not pass through the geocenter, unless the point is at a pole or on the equator. For these reasons and others, it is often convenient to bring in the coordinate system that was presented at the beginning of this chapter, the Cartesian coordinate system. However, this time it is used in its three-dimensional form, as shown in Figure 1.15.

A three-dimensional Cartesian coordinate system can be built with its origin at the center of mass of the Earth. The third coordinate, the z-coordinate, is added to the x- and y-coordinates, which are both in the plane of the equator. This system can be and is used to describe points on the surface of an ellipsoidal model of the Earth, on the actual surface of the Earth, or satellites orbiting the Earth. This system is sometimes known as the *Earth-Centered Earth-Fixed* (ECEF) coordinate system (more about that in Chapter 2).

In Figure 1.16, the relationship between the three-dimensional Cartesian coordinates of two points, ρ_1 and ρ_2, and their three-dimensional polar

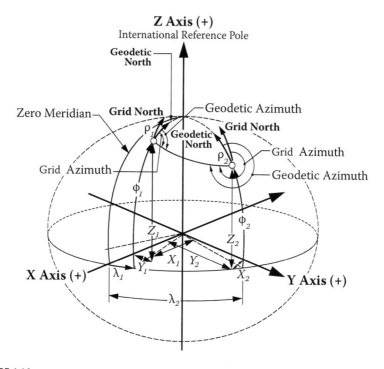

FIGURE 1.16

A few fundamentals regarding the relationship between three-dimensional Cartesian coordinates and their corresponding polar coordinates.

coordinates are illustrated on a reference ellipsoid. Under the circumstances, the polar coordinates are geocentric latitudes and longitudes. The basic relationship between the geodetic and grid azimuths is also shown in the figure. These elements outline a few of the fundamental ideas involved in commonly used coordinate systems on the Earth. Subsequent chapters will expand on these basics.

Exercises

1. Which of the expressions below is the equal to 243.1252326°?

   a. 270.1391473 grads

   b. 218.8127093 grads

   c. 4.24115 radians

   d. 243° 08′ 30″

2. Which of the following statements about two-dimensional Cartesian coordinates is incorrect?

   a. Universal Transverse Mercator coordinates are Cartesian coordinates.

   b. Most Cartesian coordinate systems are designed to place all coordinates in the first quadrant.

   c. Cartesian coordinates derived from positions on the Earth always include distortion.

   d. Directions in Cartesian coordinate systems are always reckoned from north.

3. What is the clockwise angle between the bearings N 27° 32′ 01.34″ W and S 15° 51′ 06.12″ E?

   a. 191° 40′ 55.22″

   b. 168° 19′ 04.78″

   c. 43° 23′ 07.46″

   d. 181° 04′ 22.41″

4. It is frequently necessary to convert a latitude or longitude given in degrees and decimals of degrees to degrees, minutes, and seconds and vice versa. Which of the values below correctly reflects the latitude and longitude of Youghall in degrees and decimals of degrees? Its coordinates in degrees, minutes, and seconds are 40° 25′ 33.3926″ N and 108° 45′ 57.783″ W.

   a. +40.1512021, −108.2720802

   b. +40.4259424, −108.7660508

   c. +40.42583926, −108.7658783

   d. +40.15123926, −108.2720783

5. Which of the following correctly describes a characteristic both Cartesian coordinates and polar coordinates share?

  a. Each point has only one unique coordinate pair.

  b. Coordinates are expressed in ordered pairs.

  c. Angles are measured clockwise from north in degrees, minutes, and seconds.

  d. Coordinates are always positive.

6. As one proceeds northward from the equator, which of the following does not happen?

  a. Meridians converge.

  b. Latitudinal lines are parallel.

  c. The force of gravity increases.

  d. The distance represented by a degree of latitude gets shorter.

7. Presuming that all numbers to the right of the decimal are significant, which value below is nearest to the precision of the following coordinates?

$$\varphi = 60° \ 14' \ 15.3278'' \ N$$
$$\lambda = 149° \ 54' \ 11.1457'' \ W$$

  a. ±10.0 ft

  b. ±1.0 ft

  c. ±0.10 ft

  d. ±0.01 ft

8. Which of the following statements about latitude is not true?

  a. The geocentric latitude of a point is usually smaller than the geodetic latitude of the same point.

  b. Geodetic latitude is not derived from direct measurement with optical instruments.

  c. The astronomic latitude is usually quite close to the geodetic latitude of the same point.

  d. The geodetic latitude of a point remains constant despite datum shifts.

9. Which of the following statements concerning the deflection of the vertical is correct?

   a. The deflection of the vertical is comprised of the north-south and east-west components of the difference between the geodetic and geocentric latitude and longitude of a point.

   b. The deflection of the vertical at a point can be derived from astronomic observations alone.

   c. The deflection of the vertical at a point can be derived from geodetic coordinates alone.

   d. The deflection of the vertical is used in the conversion of astronomic coordinates and astronomic directions to their geodetic counterparts and vice versa.

10. Given the astronomical coordinates,

    $\Phi = 39° 59' 38.66'' N$

    $\Lambda = 104° 59' 51.66'' W$

    the geodetic coordinates of the same point,

    $\varphi = 39° 59' 36.54'' N$

    $\lambda = 104° 59' 38.24'' W$

    and an astronomic azimuth of 352° 21′ 14.8″. What is the corresponding geodetic azimuth?

    a. 352° 21′ 06.2″

    b. 352° 21′ 12.7″

    c. 352° 21′ 01.4″

    d. 352° 21′ 23.4″

Explanations and Answers

1. Explanation:

Since there are 400 grads in a full circle of 360°, each grad is equal to 9/10ths of a degree.

$$\frac{360°}{400\,\text{grads}} = 0.90 \text{ grads in a degree}$$

Therefore, it follows that 243.1252326° divided by 0.90 will yield the measurement in grads.

$$\frac{243.125326°}{0.90 \text{ grads in a degree}} = 270.1391473 \text{ grads}$$

The decimal degrees converted to degrees, minutes, and seconds equal 243° 07′ 30.838″, and the same value can be expressed as 4.24336 radians.

Answer: **(a)**

2. Explanation:

Universal Transverse Mercator coordinates, state plane coordinates, and many other commonly used systems are indeed two-dimensional and Cartesian. It is desirable and typical to design a flat Cartesian-coordinate system to ensure that all coordinates are positive, in other words, that all are in the first quadrant. It is not possible to represent the curved Earth on a two-dimensional flat surface without some distortion. Many disciplines use Cartesian coordinates, and many do not measure directions from the north, or *y*, axis.

Answer: **(d)**

3. Explanation:

A 180° angle extends clockwise from N 27° 32′ 01.34″ W to S 27° 32′ 01.34″ E. The remaining clockwise angle between the bearings can be found by subtracting 15° 51′ 06.12″ from 27° 32′ 01.34″:

$$\begin{array}{r} 27°32'01.34 \\ -15°51'06.12 \\ \hline 11°40'55.22'' \end{array}$$

The result, added to 180°, is the clockwise angle between N 27° 32′ 01.34″ W and S 15° 51′ 06.12″ E.

$$
\begin{array}{r}
180°00′00.00″ \\
+ 11°40′55.22″ \\
\hline
191°40′55.22″
\end{array}
$$

Answer: **(a)**

4. Explanation:

40° 25′ 33.3926″ North latitude is converted to degrees and decimal degrees by

+{40 degrees + [25 min × (1 degree/60 min)] + [33.3926 sec ×
(1 min/60 sec) × (1 degree/60 min)]} = +40.4259424

108° 45′ 57.783″ West longitude is converted to degrees and decimal degrees by

−{108 degrees + [45 min × (1 degree/60 min)] + [57.783 sec ×
(1 min/60 sec) × (1 degree/60 min)]} = −108.7660508

Answer: **(b)**

5. Explanation:

In the system of Cartesian coordinates, each point has a single unique coordinate; any change in the northing, y, or easting, x, and the coordinate refers to another point entirely. In the system of polar coordinates, the vectorial angle, ρ, and the radius vector, θ, can be expressed in a number of ways and still describe the same point. A direction, expressed in an angle, may comprise half of a polar coordinate, whether it is measured in grads, radians, decimal degrees, or degrees, minutes, and seconds. While it is desirable to place the origin of a Cartesian coordinate system far to the southwest of the area of interest and thereby keep all of the coordinates positive, it is not always the arrangement. Negative coordinates are valid in both polar and Cartesian systems. However, in both systems, coordinates are given in ordered pairs.

Answer: **(b)**

6. Explanation:

As Richter discovered, the force of gravity is actually greater in the higher latitudes and decreases as one travels southward. As Newton predicted, the Earth is thicker at the equator. If the Earth were a sphere, meridians would still converge, latitudinal lines would still be parallel, and the distance represented by degree of latitude would be constant. However, the Earth is an oblate spheroid, and the length of a degree of latitude actually increases slightly as one proceeds northward due to the flattening. It is this fact that originally convinced scientists that Newton's model for the Earth was correct.

Answer: **(d)**

7. Explanation:

At approximately 60° N latitude, the length of a degree of longitude is approximately 34.5 miles. Since there are 3600 sec of latitude and longitude in 1 degree, an estimation of the length on a second of longitude at that latitude can be obtained by dividing 34.5 miles by 3600. It might be best to first convert 34.5 miles to feet.

$$34.5 \text{ miles} \times 5280 \text{ ft/mile} = 182{,}160 \text{ ft}$$

$$\frac{182{,}160 \text{ ft}}{3600} = 50.6 \text{ ft}$$

The longitude is shown to one ten-thousandth of a degree, so an estimation of the accuracy of the expression can be made by dividing the approximate length of 1 sec by 10,000.

$$\frac{50.6 \text{ ft}}{10{,}000} = 0.005 \text{ ft}$$

The same sort of approximation can be done using the latitude. The length of a degree of latitude is approximately 69 miles. To estimate the length of a second of latitude, divide 69 miles, in feet, by 3600.

$$69 \text{ miles} \times 5280 \text{ ft} = 316{,}800 \text{ ft}$$

$$\frac{316{,}800 \text{ ft}}{3600} = 101.2 \text{ ft}$$

Then, to estimate the length of 1/10,000 of a second, divide by 10,000.

$$\frac{101.2 \text{ ft}}{10,000} = 0.01 \text{ ft}$$

Answer: **(d)**

8. Explanation:

It is true that the geocentric latitude of a point is usually smaller than the geodetic latitude of the same point. Instruments set on the Earth are not oriented to the ellipsoid; they are oriented to gravity, and since geodetic latitude is on the ellipsoid, it is not feasible to measure it directly. Astronomic latitude departs from geodetic latitude per the deflection of the vertical, usually a very small amount. Further, since the datum for geodetic latitude is an ellipsoidal model of the Earth, shifts in that datum change the latitude, as was illustrated at the beginning of the chapter.

Answer: **(d)**

9. Explanation:

The deflection of the vertical is pertinent to the differences between astronomic and geodetic coordinates and directions. It has no part to play regarding geocentric information. Calculation of the deflection of the vertical requires both astronomic and geodetic information. It cannot be revealed without some aspect of both.

Answer: **(d)**

10. Explanation:

$\alpha_A - \alpha_G = +(\Lambda - \lambda)\sin \varphi$

$\alpha_A - \alpha_G = +(104° \ 59' \ 51.66'' - 104° \ 59' \ 38.24'')\sin 39° \ 59' \ 36.54''$

$\alpha_A - \alpha_G = +(13.42'')\sin 39° \ 59' \ 36.54''$

$\alpha_A - \alpha_G = +(13.42'')0.6427$

$\alpha_A - \alpha_G = +8.6''$

where:

α_A = astronomic azimuth

α_G = geodetic azimuth

Λ = astronomic longitude

λ = geodetic longitude

φ = geodetic latitude

Therefore, given the astronomic azimuth $\alpha_A = 352°\ 21'\ 14.8''$:

$\alpha_A - \alpha_G = +8.6''$

$352°\ 21'\ 14.8'' - \alpha_G = +8.6''$

$352°\ 21'\ 14.8'' - 8.6'' = \alpha_G$

$352°\ 21'\ 06.2'' = \alpha_G$

Answer: **(a)**

2

Building a Coordinate System

The actual surface of the Earth is not very cooperative. It's bumpy. There is not one nice, smooth figure that will fit it perfectly. It does resemble an ellipsoid somewhat, but an ellipsoid that fits Europe may not work for North America. And one applied to North America may not be suitable for other parts of the planet. That's why, in the past, several ellipsoids were invented to model the Earth. There are about 50 or so still in regular use for various regions of the Earth. They have been, and to a large degree still are, the foundation of coordinate systems around the world. But things are changing. And many of the changes have been perpetrated by advancements in measurement. In other words, we have a much better idea of what the Earth actually looks like today than ever before, and that has made quite a difference.

Legacy Geodetic Surveying

In measuring the Earth, accuracy unimagined until recent decades has become available from the Global Positioning System (GPS) and other satellite technologies. These advancements have, among other things, reduced the application of some geodetic measurement methods of previous generations. For example, land measurement by triangulation, once the preferred approach in geodetic surveying of nations across the globe, has lessened dramatically, even though coordinates derived from it are still relevant.

Triangulation was the primary surveying technique used to extend networks of established points across vast areas. It also provided information for the subsequent fixing of coordinates for new stations. The method relied heavily on the accurate measurement of the angles between the sides of large triangles. It was the dominant method because angular measurement has always been relatively simple compared to the measurement of the distances.

In the eighteenth and nineteenth centuries, before GPS, before the electronic distance measurement (EDM) device, and even before invar tapes, the measurement of long distances, now virtually instantaneous, could take years. It was convenient then that triangulation kept the direct measurement of the sides of the triangles to a minimum. From just a few measured baselines, a whole chain of braced quadrilaterals could be constructed (see Figure 2.1).

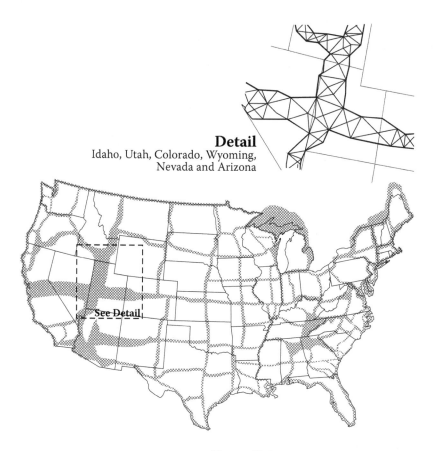

Detail
Idaho, Utah, Colorado, Wyoming,
Nevada and Arizona

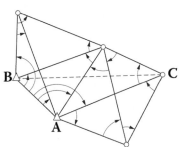

Known Data:
Length of base line AB.
Latitude and longitude of points A and B.
Azimuth of line AB.

Measured Data:
Angles to the new control points.

Computed Data:
Latitude and longitude of point C, and other
new points.
Length and azimuth of line AC.
Length and azimuth of all other lines.

FIGURE 2.1
U.S. triangulation control network map.

These quadrilaterals were made of four triangles each and could cover great areas efficiently, with the vast majority of measurements being angular.

With the quadrilaterals arranged such that all vertices were intervisible, the length of each leg could be verified from independently measured angles instead of laborious distance measurement along the ground. And when the measurements were completed, the quadrilaterals could be adjusted by least squares. This approach was used to measure thousands and thousands of chains of quadrilaterals, and these data sets are the foundations on which geodesists calculated the parameters of ellipsoids now used as the reference frames for mapping around the world.

Ellipsoids

Each ellipsoid has a name, often the name of the geodesist who originally calculated and published the figure, accompanied by the year in which it was established or revised. For example, Alexander R. Clarke used the shape of the Earth he calculated from surveying measurements in France, England, South Africa, Peru, and Lapland, including M. Struve's work in Russia and Colonel Everest's in India, to establish his Clarke 1866 ellipsoid. And even though Clarke never actually visited the United States, that ellipsoid became the standard reference model for North American Datum 1927 (NAD27) during most of the twentieth century. Despite the familiarity of Clarke's 1866 ellipsoid, it is important to specify the year when discussing it, which is true of many ellipsoids. The same British geodesist is also known for his ellipsoids of 1858 and 1880. And these are only a few of the reference ellipsoids out there.

Supplementing this variety of regional reference ellipsoids are the new ellipsoids with wider scope, such as the Geodetic Reference System 1980 (GRS80). It was adopted by the International Association of Geodesy (IAG) during the General Assembly 1979 as a reference ellipsoid appropriate for worldwide coverage. But as a practical matter, such steps do not render regional ellipsoids irrelevant any more than GPS measurements make it possible to ignore the coordinates derived from classical triangulation surveys. Any successful GIS (geographic information system) requires a merging of old and new data, and an understanding of legacy coordinate systems is, therefore, essential.

It is also important to remember that, while ellipsoidal models provide the reference for geodetic datums, they are not the datums themselves. They contribute to the datum's definition. For example, the figure for the OSGB36 datum in Great Britain is the Airy 1830 reference ellipsoid, just as the figure for the NAD83 datum in the United States is the GRS80 ellipsoid. The reference ellipsoid for the European Datum 1950 is International 1924. The reference ellipsoid for the German DHDN datum is Bessel 1841. And just to make it more interesting, there are several cases where an ellipsoid was used for more than one regional datum; for example, the GRS67 ellipsoid was the foundation for both the Australian Geodetic Datum 1966 (now superseded by GDA94) and the South American Datum 1969.

Ellipsoid Definition

To elaborate on the distinction between ellipsoids and datums, it might help to take a look at the way geodesists have defined ellipsoids. It has always been quite easy to define the size and shape of a *biaxial* ellipsoid, i.e., an ellipsoid with two axes. At least, it is easy after the hard work is done, that is, once there are enough actual surveying measurements available to define the shape of the Earth across a substantial part of its surface. Two geometric specifications will do it.

The size is usually defined by stating the distance from the center to the ellipsoid's equator. This number is known as the semi-major axis, and is usually symbolized by a (see Figure 2.2).

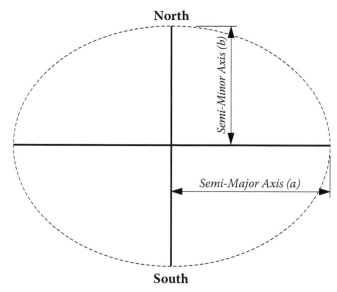

N - S *The axis of revolution for generating the ellipsoid*

flattening $f = 1 - \dfrac{b}{a}$

eccentricity $e^2 = 2f - f^2$

a - *half of the major axis the semi-major axis*

b - *half of the minor axis the semi-minor axis*

FIGURE 2.2
Parameters of a biaxial ellipsoid.

The shape can be described by one of several values. One is the distance from the center of the ellipsoid to one of its poles. That is known as the semi-minor axis, symbolized by b. Another parameter that can be used to describe the shape of an ellipsoid is the first eccentricity, or e. And finally a ratio called flattening, f, will also do the job of codifying the shape of a specific ellipsoid. Sometimes its reciprocal is used instead.

The definition of an ellipsoid then is accomplished with two numbers. It usually includes the semi-major axis and one of the other mentioned values. For example, here are some pairs of constants that are usual; first, the semi-major and semi-minor axes in meters; second, the semi-major axis in meters with the flattening, or its reciprocal; and third, the semi-major axis and the eccentricity.

Using the first method of specification, the semi-major and semi-minor axes in meters for the Airy 1830 ellipsoid are 6,377,563.396 m and 6,356,256.910 m, respectively. The first and larger number is the equatorial radius. The second is the polar radius. The difference between them, 21,307.05 m, is equivalent to about 13 miles, not much across an entire planet.

Ellipsoids can also be precisely defined by their semi-major axis and flattening. One way to express the relationship is the formula

$$f = 1 - \frac{b}{a}$$

where
  f = flattening
  a = semi-major axis
  b = semi-minor axis

The flattening for Airy 1830 is calculated as follows:

$$f = 1 - \frac{b}{a}$$

$$f = 1 - \frac{6,356,256.910m}{6,377,563.396m}$$

$$f = 1 - 0.996659150$$

$$f = \frac{1}{299.3249646}$$

In many applications, some form of eccentricity is used rather than flattening. In a biaxial ellipsoid (an ellipsoid with two axes), the eccentricity expresses the extent to which a section containing the semi-major and semi-minor axes deviates from a circle. It can be calculated as follows:

$$e^2 = 2f - f^2$$

where
   f = flattening
   e = eccentricity

The eccentricity, also known as the first eccentricity, for Airy 1830 is calculated:

$$e^2 = 2f - f^2$$

$$e^2 = 2(0.0033408506) - 0.0033408506)^2$$

$$e^2 = 0.0066705397616$$

$$e = 0.0816733724$$

Figure 2.2 illustrates the plane figure of an ellipse with two axes that is not yet imagined as a solid ellipsoid. To generate the solid ellipsoid that is actually used to model the Earth, the plane figure is rotated around the shorter axis of the two, the polar axis. The result is illustrated in Figure 2.3, where the length of the semi-major axis is the same all along the figure's equator. This sort of ellipsoid is known as an *ellipsoid of revolution*.

The length of the semi-major axis is not constant in *triaxial* ellipsoids, which are also used as models for the Earth. This idea has been around a long time. Captain A. R. Clarke wrote the following to the Royal Astronomical Society in 1860: "The earth is not exactly an ellipsoid of revolution. The equator itself

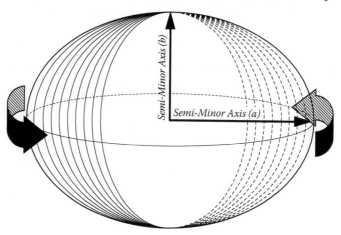

FIGURE 2.3
Biaxial ellipsoid model of the Earth.

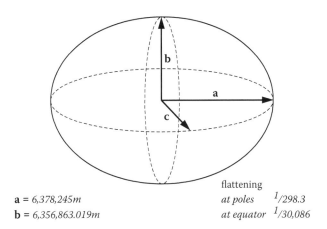

a = 6,378,245m

b = 6,356,863.019m

flattening

at poles $^1/298.3$

at equator $^1/30,086$

FIGURE 2.4
Krassovsky triaxial ellipsoid.

is slightly elliptic." Therefore, a triaxial ellipsoid has three axes with flattening at both the poles and the equator, so that the length of the semi-major axis varies along the equator, as seen in Figure 2.4. For example, the Krassovsky (1940), aka Krasovski (1940), ellipsoid is used in most of the nations formerly within the USSR.

Its semi-major axis, *a*, is 6,378,245 m, with a flattening at the poles of 1/298.3. Its semi-minor axis, *b*, is 6,356,863.019 m, with a flattening along the equator of 1/30,086. On a triaxial ellipsoid, there are two eccentricities, the meridional and the equatorial. The eccentricity, the deviation from a circle, of the ellipse formed by a section containing both the semi-major and the semi-minor axes is the meridional eccentricity. The eccentricity of the ellipse perpendicular to the semi-minor axis and containing the center of the ellipsoid is the equatorial eccentricity.

Ellipsoid Orientation

Assigning two parameters to define a reference ellipsoid is not difficult, but defining the orientation of the model in relation to the actual Earth is not so straightforward. And this is an important detail. After all, the attachment of an ellipsoidal model to the Earth makes it possible for an ellipsoid to be a geodetic datum. And the geodetic datum can, in turn, become a *terrestrial reference system* once it has actual physical stations of known coordinates easily available to users of the system.

Connection to the real Earth destroys the abstract, perfect, and errorless conventions of the original datum. They get suddenly messy because not only is the Earth's actual shape too irregular to be exactly represented by such a simple mathematical figure like an ellipsoid, but the Earth's poles wander, its surface shifts, and even the most advanced measurement methods are not perfect.

The Initial Point

When it comes to fixing an ellipsoid to the Earth, there are two methods: the old way and the new way. In the past, the creation of a geodetic datum included fixing the regional reference ellipsoid to a single point on the Earth's surface. It is good to note that the point is on the surface. The approach was to attach the ellipsoid best suited to a region at this *initial point*.

Initial points were often chosen at the site of an astronomical observatory, since observatory coordinates were usually well known and long established. The initial point required a known latitude and longitude. Observatories were also convenient places from which to determine an azimuth from the initial point to another reference point, a further prerequisite for the ellipsoid's orientation. These parameters, along with the already-mentioned two dimensions of the ellipsoid itself, made five in all. Five parameters were adequate to define a geodetic datum in this approach. The evolution of NAD27 followed this line.

The New England Datum 1879 was the first geodetic datum of this type in the United States. The reference ellipsoid was Clarke 1866, mentioned earlier, with a semi-major axis, a, of 6378.2064 km and a flattening, f, of 1/294.9786982. The initial point chosen for the New England Datum was a station known as Principio in Maryland, near the center of the region of primary concern at the time. The dimensions of the ellipsoid were defined, Principio's latitude and longitude along with the azimuth from Principio to station Turkey Point were both derived from astronomic observations, and the datum was oriented to the Earth by five parameters.

Then successful surveying of the first transcontinental arc of triangulation in 1899 connected it to the surveys on the Pacific coast. Other work tied in surveying near the Gulf of Mexico, and the system was much extended to the south and the west. It was officially renamed the United States Standard Datum in 1901.

A new initial point at Meade's Ranch in Kansas eventually replaced Principio. An azimuth was measured from this new initial point to station Waldo, because even though the Clarke 1866 ellipsoid fits North America very well, it does not conform perfectly. As the scope of triangulation across the country grew, the new initial point was chosen near the center of the continental United States to best distribute the inevitable distortion.

Five Parameters

When Canada and Mexico agreed to incorporate their control networks into the United States Standard Datum, the name was changed again to North American Datum 1913. Further adjustments were required because of the constantly increasing number of surveying measurements. This growth and readjustment eventually led to the establishment of the North American Datum 1927 (NAD27).

Before, during, and for some time after this period, the five constants mentioned were considered sufficient to define the datum. The latitude and longitude of the initial point were two. For NAD27, the latitude of 39° 13' 26.686" Nφ and longitude of 98° 32' 30.506" Wλ were specified as the coordinates of the Meade's Ranch initial point. The next two parameters described the ellipsoid itself. For the Clarke 1866 ellipsoid, these are a semi-major axis of 6,378,206.4 m and a semi-minor axis of 6,356,583.6 m. That makes four parameters. And finally, an azimuth from the initial point to a reference point for orientation was needed. The azimuth from Meade's Ranch to station Waldo was fixed at 75° 28' 09.64". Together, these five values were enough to orient the Clarke 1866 ellipsoid to the Earth and fully define the NAD27 datum.

Still other values were sometimes added to the five minimum parameters during the same era, for example, the *geoidal height* of the initial point. The assumption was sometimes made that the minor axis of the ellipsoid was parallel to the rotational axis of the Earth. The deflection of the vertical at the initial point was also sometimes considered. For the definition of NAD27, both the geoidal height and the deflection of the vertical were assumed to be zero. That meant it was often assumed that, for all practical purposes, the ellipsoid and what was known as *mean sea level* were substantially the same. As measurement has become more sophisticated, that assumption has been abandoned.

In any case, once the initial point and directions were fixed, the whole orientation of NAD27 was established. And following a major readjustment, completed in the early 1930s, it was named the North American Datum 1927.

This old approach made sense before satellite data was available. The center of the Clarke 1866 ellipsoid as utilized in NAD27 was thought to reside somewhere around the center of mass of the Earth, but the real concern had been the initial point on the surface of the Earth, not its center. As it worked out, the center of the NAD27 reference ellipsoid and the center of the Earth are more than 100 m apart. In other words NAD27, like most old regional datums, is not geocentric. This was hardly a drawback in the early twentieth century, but today truly *geocentric* datums are the goal. The new approach is to make modern datums as nearly geocentric as possible.

Geocentric refers to the center of the Earth, of course, but more particularly, it means that the center of an ellipsoid and the center of mass of the planet are as nearly coincident as possible. It is fairly well agreed that the best datums for modern applications should be geocentric and that they should have worldwide rather than regional coverage. These two ideas are due, in large measure, to the fact that satellites orbit with the center of mass of the Earth at one focus of the elliptical paths they follow. And as mentioned earlier, it is also pertinent that coordinates are now routinely derived from measurements made by the same satellite-based systems, like GPS. These developments are the impetus for many of the changes in geodesy and have made a geocentric datum an eminently practical idea. And so it has happened that satellites and the coordinates derived from them provide the raw material for the *realization* of modern datums.

Realization of a Geodetic Datum

The concrete manifestation of a datum is known as its *realization*. The realization of a datum involves the actual marking and collection of coordinates on stations throughout the region covered by the datum. In other words, the creation of the physical network of reference points on the actual Earth is part of the process of datum realization. A realized datum is ready to go to work.

For example, the users of NAD27 could hardly have begun all their surveys from the datum's initial point in central Kansas. So the Coast and Geodetic Survey (the forerunner of the National Geodetic Survey of today), as did mapping organizations around the world, produced high-quality surveys that established a network of points originally monumented by small marks in bronze disks set in concrete or rock throughout the country. These disks, their coordinates, and other attendant data became the realization of the datum, its transformation from an abstract idea into something real and usable. This same process continues today, and it contributes to a datum's maturation and evolution. Just as the surveying of chains of quadrilaterals measured by classic triangulation represented the realization of the New England Datum 1879, as the measurements grew in number and quality, they drove the evolution of that datum to become NAD27. Surveying and the subsequent setting and coordination of stations on the Earth continue to contribute to the maturation of geodetic datums today.

Terrestrial Reference Frame

The stations on the Earth's surface with known coordinates are sometimes known collectively as a *terrestrial reference frame* (*TRF*). They allow users to do real work in the real world, so it is important that they be easily accessible and their coordinate values published or otherwise easily known.

It is also important to note that there is a difference between a datum and a TRF. As stated earlier, a datum is errorless. A terrestrial reference frame is certainly not. A TRF is built from coordinates derived from actual surveying measurements. Actual measurements contain errors, always. Therefore, the coordinates that make up a TRF contain errors, however small. Datums do not. A datum is a set of constants with which a coordinate system can be abstractly defined, not the coordinated network of monumented reference stations themselves that embody the realization of the datum.

However, instead of speaking of TRFs as separate and distinct from the datums on which they rely, the word *datum* is often used to describe both the framework, which is the datum, and the coordinated points themselves, the TRF. Avoiding this could prevent a good deal of misunderstanding. For example, the relationship between two datums can be defined without ambiguity by comparing the exact parameters of each, much like comparing two ellipsoids. If one were to look at the respective semi-major axes and flattening of two biaxial ellipsoids, the difference between them would be as clear and concise as the numbers themselves. It is easy to express such differences in absolute terms. Unfortunately, such straightforward comparison is seldom the important question in day-to-day work.

On the other hand, transforming coordinates from two separate and distinctly different TRFs that both purport to represent exactly the same station on the Earth into one or the other system is an almost daily concern. In other words, it is very likely one could have an immediate need for coordinates of stations published per NAD27 expressed in coordinates in terms of NAD83. But it is unlikely one would need to know the difference in the sizes of the Clarke 1866 ellipsoid and the GRS80 ellipsoid or their orientation to the Earth. The latter is really the difference between the datums, but the coordinates speak to the relationship between the TRFs, the realization of the datums. The relationship between the datums is easily defined; the relationship between the TRFs is much more problematic. A TRF cannot be a perfect manifestation of the datum on which it lies.

The quality of the measurement technology has changed and improved with the advent of satellite geodesy. And since measurement technology, surveying techniques, and geodetic datums evolve together, so datums have grown in scope to worldwide coverage, improved in accuracy, and become as geocentric as possible.

In the past, the vast majority of coordinates involved would be determined by classical surveying, as described previously. Originally, triangulation work was done with theodolites, towers, and tapes. The measurements were Earth-bound, and the resulting stations were solidly anchored to the ground too, like the thousands of Ordnance Survey triangulation pillars on British hilltops and the million or more bronze disks set across the United States. These terrestrial reference frames provide users with accessible, stable references so that positioning work can commence from them.

New Geocentric Datum

The relationship between the centers of reference ellipsoids and the Earth's center was not an important consideration before space-based geodesy. Regional reference ellipsoids were the rule.

Even after the advent of the first electronic distance-measurement devices, the general approach to surveying still involved the determination of horizontal coordinates by measuring from point to point on the Earth's surface and adding heights, otherwise known as elevations, separately. So while the horizontal coordinates of a particular station would end up on the ellipsoid, the elevation, or height, would not. In the past, the precise definition of the details of this situation was not really an overriding concern. Because the horizontal and vertical coordinates of a station were derived from different operations, they lay on different surfaces; whether the datum was truly geocentric was not really pertinent. One consequence of this approach is that the polar and equatorial axes of older, nongeocentric ellipsoids do not coincide with the polar axis and equatorial plane of the actual Earth. The axis of the ellipsoid and the axis of the Earth were often assumed to be parallel and within a few hundred meters of each other, but not coincident, as shown in Figure 2.5.

Over the last decades, two objectives have emerged: (a) ellipsoidal models that represent the entire Earth, not just regions of it, and (b) fixing such an ellipsoid very closely to the center of mass of the planet rather than an arbitrary initial point on the surface. A large part of the impetus was eminently practical. The change was necessary because the NAD27 terrestrial reference frame simply could not support the dramatically improved measurement technology. The accuracy of its coordinates was just not as good as the surveying work the users of the datum were doing.

In the old datum, surveyors would begin their measurements and calculations from an established station with published NAD27 coordinates. They would then move on to set a completely new project point that they required. Once that was done, they would check their work. This was done by pushing on to yet another, different, known station that also had a NAD27 published

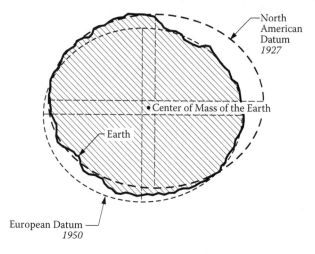

FIGURE 2.5
Regional ellipsoids.

coordinate. Unfortunately, this checking frequently revealed that their new work had created coordinates that simply did not fit in with the published coordinates at the known stations. They were often different by a considerable amount. Under such circumstances, the surveyors had no choice but to adjust the surveyed measurements to match the published coordinates. New work had to fit into the existing framework of the national network of coordinates.

Satellite positioning, and more specifically GPS, made it clear that the accuracy of surveying had made a qualitative leap. It was also apparent that adjusting satellite-derived measurements to fit the less accurate coordinates available from NAD27 was untenable. A new datum was needed: a datum that oriented to the geocenter like the orbits of the satellites themselves; a datum that could support a three-dimensional Cartesian coordinate system and thereby contribute to clearly defining both the horizontal and vertical aspects of the new coordinates.

So the North American Datum of 1983 replaced the North American Datum of 1927. The new datum was fundamentally different. With the advent of space geodesy—such as *Satellite Laser Ranging (SLR), Lunar Laser Ranging (LLR), Very Long Baseline Interferometry (VLBI), Doppler Orbitography by Radiopositioning Integrated on Satellite (DORIS), and the Global Positioning System (GPS)*—tools became available to connect points and accurately determine coordinates on one global reference surface. Of the many space-based techniques that emerged in the 1980s and matured in the 1990s, GPS is of particular importance. The receivers are relatively small, cheap, and easy to operate. And the millimeter-to-centimeter level of positioning accuracy has been widely demonstrated over long baselines. Nevertheless, very few GPS observations were used in the establishment of NAD83.

It took more than 10 years to readjust and redefine the horizontal coordinate system of North America into NAD83. More than 1.7 million weighted classical surveying observations were involved, along with some 30,000 EDM-measured baselines, 5,000 astronomic azimuths, 655 Doppler stations positioned using the TRANSIT satellite system, and 112 very-long-baseline interferometry (VLBI) vectors. In short, the North American Datum of 1983 (NAD83) can be said to be the first civilian coordinate system established using satellite positioning. And it was much more accurate than NAD27.

So when NAD83 coordinates were implemented across the United States, coordinates shifted. Across a small area, the coordinate shift between the two datums is almost constant, and in some areas the shift is slight. In fact, the smallest differences occur in the middle of the United States. However, as the area considered grows, one can see that there is a significant, systematic variation between NAD27 coordinates and NAD83 coordinates. The differences can grow from about −0.7″ to +1.5″ in latitude, which is up to almost 50 m north-south. The change between NAD27 and NAD83 coordinates is generally larger east-west, from −2.0″ to about +5.0″ in longitude, which means the maximum differences can be over 100 m in that direction. The longitudinal shifts are actually a bit larger than that in Alaska, ranging up to 12.0″ in longitude.

It is important to note that if the switch from NAD27 to NAD83 had only involved a change in surveying measurements made on the same ellipsoid, the changes in the coordinates would not have been that large. For example, had NAD83 coordinates been derived from satellite observations and then been projected onto the same Clarke 1866 ellipsoid used for NAD27, the change in coordinates would have been smaller. But, in fact, the center of the ellipsoid shifted approximately 236 m from the nongeocentric Clarke 1866 ellipsoid to the geocentric GRS80 ellipsoid.

The evolution of the new datum has continued. NAD83 was actually in place before GPS was operational. As GPS measurements became more common, they turned out to be more accurate than the coordinates assigned to the network of control points on the ground. NAD83 needed to be refined. Frequently, states took the lead, and the U.S. National Geodetic Survey (NGS) participated in cooperative work that resulted in readjustments. The new refinements were referred to with a suffix, such as NAD83/91, and the term *High Precision GPS Network (HPGN)* was used. Today, *High Accuracy Reference Network (HARN)* is the name most often associated with these improvements of NAD83.

The *World Geodetic System 1984* (WGS84) is the geodetic reference used by GPS. WGS84 was developed for the United States Defense Mapping Agency (DMA). The agency's name was changed to National Imagery and Mapping Agency (NIMA), and today it is known as the National Geospatial-Intelligence Agency (NGA). GPS receivers compute and store coordinates in terms of WGS84. They transform them to other datums when information is displayed. WGS84 is the default for many GIS platforms as well.

The original realization of the WGS84 was based on observations of the TRANSIT satellite system. These positions had 1- to 2-m accuracy, but over the years, the realizations have improved. It should be noted that WGS84's ellipsoid and the GRS80 ellipsoid are very similar; they both use biaxial reference ellipsoids with only slight differences in the flattening. WGS84 has been enhanced on several occasions to a point where it is now very closely aligned to ITRF, the International Terrestrial Reference Frame.

WGS84 has been periodically improved to account for plate tectonics. The first such enhancement was in 1994 on GPS week 730; for this purpose, GPS weeks are counted from midnight January 5, 1980. This caused the name of WGS84 to acquire a suffix. It was then known as WGS84 (G730). The next update was a couple of years later, when it became known as WGS84 (G873). The latest improvement along this line resulted in WGS84 (G1150).

It is important to note that these changes have caused WGS84 to drift farther and farther from NAD83. While it is often presumed that the WGS84 as originally rolled out was nearly the same as NAD83 (86), things have changed. In fact, the difference between a position in NAD83 (CORS96) and a position in WGS84 (G1150) can approach 1 or 2 m today. At the same time,

WGS84 has become virtually coincident with the International Terrestrial Reference Frame. WGS84 (G730) was very close to ITRF92; WGS84 (G873) was close to ITRF96; and WGS84 (G1150) is close to ITRF00. WGS84 (G1150) is also currently the reference for the GPS broadcast ephemeris.

GPS information has also contributed to bringing the center of ellipsoids very close indeed to the actual center of mass of the Earth. The geocenter is a focus of the satellites' orbits and the origin of the measurements derived from them. Coordinates derived directly from GPS observations are often expressed in three-dimensional Cartesian coordinates, X, Y, and Z, with the center of mass of the Earth as the origin.

Geocentric Three-Dimensional Cartesian Coordinates

A three-dimensional Cartesian system requires three axes and a clear definition of both their origin and their direction. If these parameters can be attached to the Earth, then every position on the planet, and in its vicinity, can have a unique three-dimensional Cartesian coordinate. However, as noted in Chapter 1, when you bring in the real world, things get messy. For example, the relationship of the surface of the Earth, its center, and even its axis of rotation is not constant and unchanging.

The Earth wobbles in motions known as *precession* and *nutation*. Precession is the long-term movement of the polar axis. It moves in a circle with a period of something approximating 25,800 years. At the moment, the planet's spin axis almost points to Polaris, but in 14,000 years or so it might point to Vega. Nutation is the movement of the Earth with a cycle of about 18.6 years, mostly attributable to the moon. The Earth's rotation rate also varies. It is a bit faster in January and slower in July. And then there is the wandering of the Earth's axis of rotation relative to the Earth's surface, called *polar motion*.

Polar motion is a consequence of the actual movement of the Earth's spin axis as it describes an irregular circle with respect to the Earth's surface. The circle described by this free Eulerian motion of the pole has a period of about 435 days or so. It takes approximately that long for the pole to complete a circle that has a diameter of about 12 to 15 m. This part of the polar motion is known as the *Chandler period*, named after American Astronomer Seth C. Chandler, who described it in papers in the *Astronomical Journal* in 1891. Another aspect of polar motion is sometimes called *polar wander*. It is about 0.004 sec of arc per year as the pole moves toward Ellesmere Island. Both aspects are shown generally in Figure 2.6.

Therefore, one can say that the Earth has a particular axis of rotation, equator, and zero meridian for an instant before they all change slightly in the next instant. Within all this motion, how do you define the origin and direction of the three needed axes for the long term? One way is to choose a moment in time and consider them fixed as they are at that instant. That was how it was done. A moment was chosen by the Bureau International de l'Heure (BIH). It was midnight on New Year's Eve 1983, or January 1, 1984

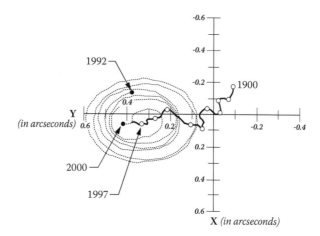

FIGURE 2.6
Polar wander from 1900 and polar motion from 1992 to 2000.

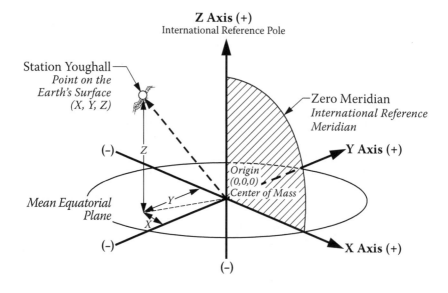

FIGURE 2.7
A three-dimensional Cartesian coordinate in the Conventional Terrestrial System.

(UTC). It is also known as an *epoch* and can be written 1984.0. So we now use the axes illustrated in Figure 2.7 as they were at that moment.

This resulting system is known as the *Conventional Terrestrial Reference System* (*CTRS*), or simply the *Conventional Terrestrial System* (*CTS*). The origin is the center of mass of the whole Earth, including oceans and atmosphere, called the geocenter. The *x*-axis is a line from that geocenter through its intersection at the zero meridian, also known as the *International Reference*

Meridian (IRM), with the internationally defined conventional equator. The *y*-axis is extended from the geocenter along a line perpendicular from the *x*-axis in the same mean equatorial plane. That means that the positive end of the *y*-axis intersects the actual Earth in the Indian Ocean. In any case, they both rotate with the Earth around the *z*-axis, a line from the geocenter through the internationally defined pole known as the *International Reference Pole* (IRP). The three-dimensional Cartesian coordinates (*x, y, z*) derived from this system are sometimes known as *Earth-Centered Earth-Fixed (ECEF)* coordinates, and this system has been utilized in NAD83, WGS 84, and ITRF (see next section). It is a *right-handed*, orthogonal system and can be described by the following model. The horizontally extended forefinger of the right hand symbolizes the positive direction of the *x*-axis. The middle finger of the same hand extended at right angles to the forefinger symbolizes the positive direction of the *y*-axis. The extended thumb of the right hand, perpendicular to them both, symbolizes the positive direction of the *z*-axis.

In this three-dimensional, right-handed coordinate system, the *x*-coordinate is a distance from the *y–z* plane measured parallel to the *x*-axis. It is always positive from the zero meridian to 90° W longitude and from the zero meridian to 90° E longitude. In the remaining 180°, the *x*-coordinate is negative.

The *y*-coordinate is a perpendicular distance from the plane of the zero meridian. It is always positive in the Eastern Hemisphere and negative in the Western Hemisphere.

The *z*-coordinate is a perpendicular distance from the plane of the equator. It is always positive in the Northern Hemisphere and negative in the Southern Hemisphere.

Here is an example, with the position of the station Youghall expressed in three-dimensional Cartesian coordinates of this type as expressed in meters, the native unit of the system:

$X = -1564831.1855$ m

$Y = -4605604.7477$ m

$Z = 4115817.6900$ m

The X-coordinate for Youghall is negative because its longitude is west of 90° W longitude. The Y-coordinate is negative because the point is west of the zero meridian. The Z-coordinate is positive because the point is in the Northern Hemisphere.

The system works well, but what about earthquakes, volcanic activity, tides, subsurface fluid withdrawal, crustal loading/unloading, and the many other forces that contribute to the continuous drift in tectonic plates? Certainly these cause the plates and the surface points on them to move relative to the geocenter and relative to each other. Hence changes in the three-dimensional Cartesian coordinates are inevitable. Not only is the entirety of the Earth's surface always in motion with respect to its center of mass, but there is also relative motion between the approximately 20 large tectonic plates that make up that surface. For example, stations on separate plates can move as much as 150 mm per year simply because the ground on which one station stands is slowly shifting in relation to the ground at the other. It was this fact, among other things, that led to the establishment of the *International Earth Rotation Service (IERS)*; the previously mentioned Bureau International de l'Heure (BIH) was its predecessor. And the IERS introduced a heretofore unheard of and remarkable aspect to coordinates, velocity.

The IERS

The International Earth Rotation Service (IERS) is an organization that began operations in Paris at the beginning of 1988 under the auspices of the International Astronomical Union and the International Union of Geodesy and Geophysics. The IERS has formally defined the International Reference Pole, the International Reference Meridian, the plane of the conventional equator, and the other components of the three-dimensional Cartesian system just described. These are part of a broader system known as the *International Terrestrial Reference System (ITRS)*.

Because work in crustal deformation and the movement of the planet's axis required an extremely accurate foundation, the IERS originally introduced the International Terrestrial Reference System (ITRS) and its first realization, the *International Terrestrial Reference Frame of 1988 (ITRF88)*. As mentioned earlier, a realization is the concrete manifestation of a datum by measurements made at points on the Earth. In the case of the ITRF, the region covered includes the whole world. Please note that the digits appended to ITRF represent the year up to which the data sets have been used in the realization. In fact, from its beginning in 1988, and nearly every year since, IERS has published a list of new and revised positions with their velocities for more than 500 stations around the world. In other words, ITRF89, ITRF90, etc., followed ITRF88. So the IERS released ITRF89, ITRF90, ITRF91, ITRF92, ITRF93, and ITRF94. There was not a release in 1995. The next were ITRF96 and ITRF97, followed by ITRF00 and ITRF05 for 2000 and 2005, respectively.

This international organization does the hard work of compiling the measurements and calculating the movement of our planet. The data needed for this work come from measurements made by the Global Positioning System (GPS), very-long-baseline interferometry (VLBI), satellite laser ranging (SLR), and satellite radio positioning (DORIS) at a set of stations around the planet that realize the ITRF. These values are utilized more and more in part because of the high quality of the data, and in part because ITRS provides an international reference system that directly addresses crustal motion at particular monumented control stations. The ITRF stations are moving and recognized to be doing so. Therefore, each is described by a position and a velocity, and every position—every set of coordinates in these realizations—refers to a station's position at a particular moment. This moment is known as the Reference Epoch (RE). For example, the RE for ITRF94 is 1993.0, or more specifically January 1, 1993, at exactly 0:00 UTC. The RE for ITRF00 is 1997.0, and the RE for ITRF05 is 2000.0. Even stations on the most stable part of the North American plate are in horizontal motion continuously at rates that range from 9 to 21 mm every year. While it must be said that these changes in coordinates are very slight indeed, they are changes nonetheless, and ITRS accounts for them.

A good deal of practical application of this system has developed. The U.S. National Geodetic Service (NGS), an office of NOAA's National Ocean Service, the arm of the U.S. federal government that defines and manages the National Spatial Reference System (NSRS), is now utilizing ITRF data. In March 2002 the NGS started to upgrade its published NAD83 positions and velocities for a portion of the national network (known as the CORS sites) to be equal to the ITRF2000 positions and velocities.

The acronym CORS stands for the network of continuously operating reference stations. GPS signals collected and archived at these sites of known position provide base data that serve as the foundation for positioning in the United States and its territories. For example, when GPS data is collected at an unknown station, it can be processed with data from a CORS station to produce positions that have centimeter-level accuracy in relation to the NSRS, and now ITRF2000 as well.

Here is an illustration of the application of ITRF data. The coordinate values for the position of station AMC2 in Colorado Springs, Colorado, are shown here as retrieved from the NGS database. First is the NAD83 (epoch 2002) position of the station in three-dimensional Cartesian coordinates as it was transformed from the ITRF00 position (epoch 1997.0) in March 2002:

$X = -1248595.534$ m

$Y = -4819429.552$ m

$Z = 3976506.046$ m

Now here is the position of the same station based upon the ITRF00 position (epoch 1997.0) as it was computed in August 2006 using 1,673 days of data.

$X = -1248596.072$ m

$Y = -4819428.218$ m

$Z = 3976506.023$ m

The question, "Where is station AMC2?" might be more correctly asked, "Where is station AMC2 now?" And, in fact, the latter question can be answered by calculating new positions for the station based on its velocities.

The location of the AMC2 station can also be stated in both three-dimensional Cartesian coordinates and in latitude, longitude, and height above the ellipsoid. Here is the AMC2 ITRF00 (1997.0) position calculated in August 2006 expressed in geographic coordinates and ellipsoidal height:

Latitude = 38° 48′ 11.24915″ N

Longitude = 104° 31′ 28.53276″ W

Ellipsoid height = 1911.393 m

Three-dimensional Cartesian coordinates and geographical coordinates with ellipsoidal heights can be converted from one to the other. Here are the expressions for deriving the three-dimensional Cartesian coordinates from latitude, longitude, and height above the ellipsoid:

$$X = (N + h) \cos \varphi \cos \lambda$$
$$Y = (N + h) \cos \varphi \sin \lambda$$
$$Z = (Nb^2/a^2 + h) \sin \varphi$$

where
   latitude = φ
   longitude = λ
   height above the ellipsoid = h
   a = semi-major axis
   b = semi-minor axis
   N = the east-west local radius of the reference ellipsoid in meters

$$N = a^2 (a^2 \cos^2 \varphi + b^2 \sin^2\varphi)^{-\frac{1}{2}}$$

Notice in Figure 2.8 that this computation requires the introduction of an ellipsoid to represent the Earth itself. This is interesting because it means that, just by looking at an *x*-, *y*-, and *z*-coordinate in this system, one cannot be certain whether the point is on the Earth's surface, deep inside it, or in outer space.

Transforming Coordinates

Transformations are mathematical mechanisms used to move coordinates from one datum to another, and there are several methods. And while it is true that most such work is left to computer applications today, it is, at best, unwise to accept the results uncritically. Therefore, here is an outline of some of the qualities of some typical datum transformation methods.

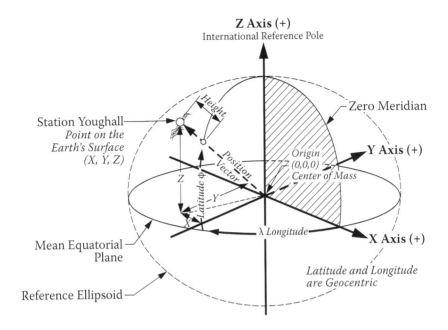

X = *-1564831.1855m*
Y = *-4605604.7477m*
Z = *4115817.6900m*

FIGURE 2.8
A position derived from space-based geodetic measurement.

First, datum transformations ought to be distinguished from coordinate conversions. Coordinate conversion is usually understood to mean the reexpression of coordinates from one form to another, but both resting on the same datum. For example, the calculation of a plane coordinate such as UTM (NAD83) from a point's expression as a geographic coordinate (NAD83) would be a coordinate conversion.

Datum transformation, on the other hand, usually means that the coordinates do in fact change from an original datum to a target datum. For example, the alteration of a geographic coordinate on one datum, e.g., latitude and longitude in NAD83 into a geographic coordinate on another datum, e.g., latitude and longitude in NAD27, would amount to a datum transformation. In this case, there is actually a change from one ellipsoid to another. A further complication is presented by the fact that the center of the ellipsoid of reference for NAD83, GRS80, and the center of the ellipsoid of reference for NAD27, Clarke 1866, do not coincide. Other typical difficulties include the orientation of the axes of the original and target datums. They may require rotation to coincide, or the scale of the distances between points on one datum may not be at the same scale as the distances on the target datum, etc.

Each of these changes can be addressed mathematically to bring integrity to the transformation of coordinates when they are moved from one system into another. In other words, the goal is to not degrade the accuracy of the coordinates in the terrestrial reference frame as they are transformed. However, datum transformation cannot improve the accuracy of those coordinates, either. For example, if the distance between point A and point B is incorrect in the original datum, it will be just as wrong when it is transformed into a target datum. The initial consistency of the coordinate network to be transformed, i.e., the accuracy with which the coordinates represent the relative positions of the actual points on the Earth, is important.

Common Points

In datum transformations, it is best if some of the points involved have been surveyed in both the original datum and the target datum. These are usually called common points. A common point has, of course, one coordinate in the original datum and an entirely different coordinate in the target datum; still, both represent the same position on the Earth.

The accuracy and the distribution of these coordinates in both the original and the target datum is an important factor in the veracity of a datum transformation. When it comes to datum transformation, the more common points there are, the better is the result. And they are best if evenly distributed through the network. These factors affect the results no less than the actual method used to perform a datum transformation.

Clearly, some of the surveyed common points in the target datum can be used after the transformation is completed to check the work. The surveyed coordinates can then be compared to the transformed coordinates to evaluate the consistency of the operation.

Molodenski Transformation

This method is named for the mid-twentieth-century Russian physicist, M. S. Molodenski. The Molodenski transformation (Figure 2.9) is sometimes known as the three-parameter or five-parameter transformation. It is used on-board some GPS receivers. In fact, the growth in the utilization of GPS has probably increased the number of computer applications relying on this method. In this context, the Molodenski transformation is often implemented to transform coordinates from WGS84 into a local projection.

The Molodenski transformation is simple in conception and available in many standard GIS software platforms as well. It rests on shifts to the three geocentric coordinates that are applied directly to geographical coordinates. It usually requires ellipsoidal parameters from the original and the target system as well as the size of the shift in the x-, y-, and z-geocentric

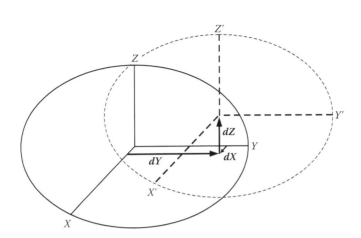

FIGURE 2.9
Molodenski transformation.

coordinates. In other words, it uses simple straightforward formulas to shift the origin from the original datum to the target datum along the x-, y-, and z-axes (ΔX, ΔY, ΔZ) based on the averaged differences between the x-, y-, and z-coordinates of the previously mentioned common points. There is no scaling or rotation in this method.

Like all such mathematical operations, the worth of this sort of transformation is dependent on the consistency of the coordinate values available, but at best it can only produce a transformation with moderate accuracy. The size of the area being transformed bears on the accuracy of the transformation.

The Molodenski transformation is based on the assumption that the axes of the original ellipsoid and the target ellipsoid are parallel to each other. That is seldom true, but if the work involves only a small area, the effect of the assumption may be insignificant. However, as the size of the area grows, so does the inaccuracy of this method of transformation. In short, the Molodenski method is satisfactory if some of the work requires modest accuracy, but rotation and scale parameters are needed for more precise work (Figure 2.10).

Seven-Parameter Transformation

This method is also known as a Helmert or Bursa-Wolf transformation. It bears remembering that datum transformations do not improve the accuracy of the coordinates they transform. They cannot do that. However, when the number of parameters considered is increased, the result is an improvement in the fit of the coordinates in the target datum.

To transform from one geocentric datum to another, one could use the seven parameters of the Bursa-Wolf approach; three translations, three rotations, and one scale factor. The sum of the x, y, and z translations accomplishes

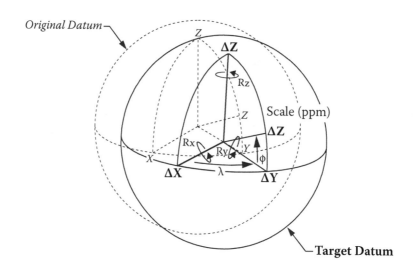

FIGURE 2.10
Translation and rotation.

the shifting of the origin so that the origins of the original datum and the target datum match. The shifts are usually expressed as ΔX, ΔY, ΔZ or DX, DY, and DZ, but they are sometimes shown as u, v, and w. In any case, the three shift distances are specified in meters. Their objective is to shift the ellipsoid along each of its three axes.

Then, through the rotation of each of the axes, the original datum and the target datum axes are made parallel to one another. The three rotation parameters of the x-, y-, and z-axes are symbolized by EX, EY, and EZ or rX, rY, rZ, though they are sometimes shown as ε_x, ε_y, ε_z or ε, ψ, and ω. The three rotation parameters specify the angles. The angles are usually less than 5 arc-sec and are calculated by producing a combined rotation matrix. Finally, the transformation is scaled. The scale factor is usually calculated in parts per million.

This method is also known as the three-dimensional Helmert, three-dimensional conformal, or three-dimensional similarity. The seven-parameter transformation should start with at least three coordinates of points that are common to the original and the target system; more are better.

Again, the quality of the results depends on the consistency of the set of common coordinated points utilized by the original and target side of the work. This transformation is available in many GIS software packages, and its accuracy is better than that available with the Molodenski transformation. The seven-parameter transformation does require heights for the common coordinated points in both the original and the target systems, and the results depend on the consistency of the coordinates in both systems.

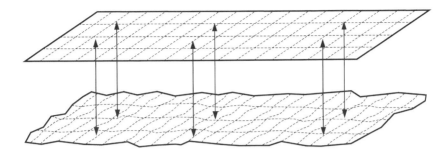

FIGURE 2.11
Surface fitting.

Surface Fitting

Surface fitting, illustrated in Figure 2.11, is also known as the transformation grid, bilinear gridded interpolation, or the grid-based interpolation method. This is the best approach to datum transformation.

The United States, Canada, and Australia use the method. In 1986, when the datum changed in the United States, it was clear that there were no constant values that could easily move the geographical coordinates from the original datum, NAD27 on the Clarke 1866 ellipsoid, to the target datum, NAD83 on the GRS80 ellipsoid. Among the reasons were the centuries of conventional surveying that introduced unavoidable inconsistencies into the original coordinates. Under the circumstances, it would have been unworkable to transform the NAD27 coordinates into NAD83 with the Molodenski method or the seven-parameter method. What was needed was an approach that was more finely tuned and specific.

The *North American Datum Conversion Utility* (*NADCON*) was developed. The program uses grids and an error-averaging strategy based on real data. When a coordinate is input for transformation, the necessary shift calculations are based on a grid expressed in a database that contains the shifts at tens of thousands of points in an extensive grid network. This grid of control points has shifts that are known, and that information is used to estimate the shifts at other locations.

This approach is based on the idea that it is possible to find the necessary shift in a coordinate to transform it from its original datum to a target datum by interpolation from known shifts for a number of control points in the same area. For example, a particular geographic coordinate to be transformed will fall within a grid cell that has points of known shifts at each of its four corners. The application of a bilinear interpolation algorithm can thereby derive the necessary shift at the given point. The interpolation uses two grid files: one for the shifts in latitude and one for the shifts in longitude. This rubber-sheeting method is good, but it requires an extensive grid database to be successful. Building such a database needs the devotion of significant resources and almost certainly the auspices of governmental agencies. It is a minimum-curvature method, and it is probably the most popular transformation routine in the United States. It is a surface-fitting type of transformation.

The software behind the surface-fitting transformation is not based on simple formulas, but it can be operated from a simple user interface that emphasizes simple shifts in latitude and longitude. Several points common to the original and the target coordinate systems are used in the surface-fitting method.

On the positive side, the surface-fitting transformation is quite accurate and often driven by a simple user interface. It is integral to many standard GIS software packages. On the other hand, it is mathematically complex and requires that the original and the target coordinate systems have many common points.

The number of common points available and the accuracy required are important considerations in choosing the appropriate transformation method. The surface-fitting or grid-shift techniques like those used in NADCON provide the best results. The Molodenski transformation provides the least accuracy.

Exercises

1. Which of the following pairs of parameters is not usually used to define a reference ellipsoid?
   a. Semi-major and semi-minor axes
   b. Semi-major axis and the reciprocal of the flattening
   c. Semi-major axis and the eccentricity
   d. Semi-major axis and the reciprocal of the eccentricity

2. Which of the following is the reference ellipsoid for NAD27?
   a. Clarke 1858
   b. Bessel 1841
   c. Clarke 1866
   d. Clarke 1880

3. Which of the following is the reference ellipsoid for North American Datum 1983?
   a. OSGB36
   b. GRS80
   c. WGS66
   d. ITRF88

4. Which of the following is not a characteristic of a conventional terrestrial reference system (CTRS)?
   a. The X-, Y-, and Z-axes are all perpendicular to one another.
   b. The Z-axis moves due to polar motion.
   c. The X- and Y-axes rotate with the Earth around the Z-axis.
   d. The origin is the center of mass of the Earth.

5. Which of the following statements about GRS80 is not correct?
   a. The GRS80 ellipsoid is slightly different from the WGS84 ellipsoid.
   b. NAD83 uses GRS80 as its reference ellipsoid.
   c. GRS80 is geocentric.
   d. The GRS80 ellipsoid is the reference ellipsoid for GPS.

6. Which statement about biaxial and triaxial ellipsoids is correct?

    a. The lengths of the semi-major and semi-minor axes are consistent in both a biaxial ellipsoid and a triaxial ellipsoid.

    b. The reference ellipsoid for NAD83 is triaxial and the reference ellipsoid for NAD27 is biaxial.

    c. The equator of a triaxial ellipsoid is elliptical; the equator of a biaxial ellipsoid is a circle.

    d. Both triaxial and biaxial ellipsoids have flattening at the equator.

7. In the Conventional Terrestrial Reference System (CTRS), with its origin at the geocenter of the Earth and a right-handed orientation of the X-, Y-, and Z-axes, the Z-coordinate is positive in the Northern Hemisphere and negative in the Southern Hemisphere. Which answer correctly states the regions of the Earth where the X-coordinate is positive and where it is negative?

    a. The X-coordinate is negative from the zero meridian to 90° W longitude and from the zero meridian to 90° E longitude. In the remaining 180° of longitude, the X-coordinate is positive.

    b. The X-coordinate is positive in the Eastern Hemisphere and negative in the Western.

    c. The X-coordinate is positive from the zero meridian to 90° W longitude and from the zero meridian to 90° E longitude. In the remaining 180°, the X-coordinate is negative.

    d. The X-coordinate is positive in the Western Hemisphere and negative in the Eastern.

8. What constitutes the realization of a geodetic datum?

    a. The assignment of a semi-major axis and another parameter, such as the first eccentricity, flattening, and/or the semi-minor axis

    b. The monumentation of the physical network of reference points on the actual Earth that is provided with known coordinates in the subject datum

    c. The definition of an initial point at a known latitude and longitude and a geodetic azimuth from there to another known position along with the two parameters of the associated ellipsoid, such as its semi-major and semi-minor axes

    d. The transformation of its coordinates into another datum using the seven parameters of the Bursa-Wolf process

9. Which of the following acronyms identifies a realization of the International Terrestrial Reference System?
   a. VLBI
   b. SLR
   c. ITRF88
   d. IERS

10. Which of these transformation methods is likely to produce the most accurate results?
    a. Molodenski
    b. Seven-parameter
    c. Surface fitting
    d. Krassovsky

Explanations and Answers

1. Explanation:

Compared with a sphere, an oblate ellipsoid is flattened at the poles. It is often expressed as a ratio. This ratio, and sometimes its reciprocal, is used as one specification for reference ellipsoids used to model the Earth. It cannot fulfill the definition alone, however, and another parameter is always needed. That is frequently the semi-major axis, i.e., the long equatorial axis. An ellipsoid can also be fully determined with the statement of the semi-major axis and the semi-minor axis, i.e., the short or polar axis. Eccentricity, or first eccentricity, is often used in conjunction with the semi-major axis as the definition of a reference ellipsoid, but the reciprocal of the eccentricity is not.

Answer: **(d)**

2. Explanation:

In the 1860s, while working as head of the Trigonometrical and Leveling Departments of the Ordnance Survey in Southampton in Britain, Captain Alexander Clarke worked as a geodesist. He made and published calculations of the size and shape of the Earth, first in 1858. Then, in 1866, Captain Clarke made his second determination of a reference ellipsoid for the Earth, and the figure became a standard reference for U.S. geodesy and the North American Datum 1927. He published another reference ellipsoid in 1880.

Answer: **(c)**

3. Explanation:

Ordnance Survey Great Britain 1936 (OSGB36) is a traditional geodetic datum or terrestrial reference frame. Its reference ellipsoid is Airy 1830. The ITRF88 was the first International Terrestrial Reference Frame, and its reference ellipsoid is Geodetic Reference System 1980 (GRS80), the same ellipsoid used for NAD83.

Answer: **(b)**

4. Explanation:

All of the answers mention aspects of a conventional terrestrial reference system (CTRS), except b. While it is true that the actual polar axis wanders, the International Reference Pole (IRP) does not move with polar motion because, by international agreement, it stands where it was on 1 January 1903.

Answer: **(b)**

5. Explanation:

The parameters of the Geodetic Reference System 1980 (GRS80), adopted by the International Association of Geodesy (IAG) during the General Assembly 1979, are:

Semi-minor axis (polar radius)	b	6356752.3141 m
First eccentricity squared	e^2	0.00669438002290
Flattening	f	1 : 298.257222101

The parameters of the World Geodetic System1984 (WGS84) ellipsoid are:

Semi-minor axis (polar radius)	b	6356752.3142 m
First eccentricity squared	e^2	0.00669437999013
Flattening	f	1 : 298.2572223563

These ellipsoids are very similar. The semi-major axis of GRS80 is 6,378,137 m, and its inverse of flattening is 298.257222101. The semi-major axis of WGS84 is the same within the accuracy shown at 6,378,137 m, but its inverse of flattening is slightly different at 298.257223563.

However, GRS80 is not the geodetic reference system used by GPS. The relevant reference system is the WGS84 ellipsoid, which was developed for the U.S. Defense Mapping Agency (DMA), now known as NIMA (National Imagery and Mapping Agency). GPS receivers compute and store coordinates in terms of WGS84, and most can then transform them to other datums when information is displayed.

Answer: **(d)**

6. Explanation:

A triaxial ellipsoid such as the Krassovsky ellipsoid has flattening at the equator, but biaxial ellipsoids do not. The equator of a triaxial ellipsoid is elliptical, but the equator of a biaxial ellipsoid is a circle, since its semi-major axis has a constant length. GRS80, the reference ellipsoid for NAD83, and the Clarke 1866 ellipsoid, the reference ellipsoid of NAD27, are both biaxial.

Answer: (c)

7. Explanation:

In the right-handed, three-dimensional coordinate system known as CTRS, the X-coordinate is positive from the zero meridian to 90° W longitude and from the zero meridian to 90° E longitude. In the remaining 180°, the X-coordinate is negative. The Y-coordinate is positive in the Eastern Hemisphere and negative in the Western. The Z-coordinate is positive in the Northern Hemisphere and negative in the Southern.

Answer: (c)

8. Explanation:

The realization of a datum is the assignment of coordinates in that datum to actual physical monumentation on the ground. Following the realization of a datum, subsequent work can be tied to physically marked points on the Earth rather than an abstract concept of the datum (its theoretical origin and axes). This concrete manifestation of a datum is ready to go to work in the real world.

Answer: (b)

9. Explanation:

The international organization that compiles the measurements and calculates the movement of our planet is the International Earth Rotation Service (IERS). The first realization of the International Terrestrial Reference System produced by the IERS was the International Terrestrial Reference Frame of 1988 (ITRF88). The IERS also defines the International Reference Pole (IRP), and the data needed for this work comes from GPS, very-long-baseline interferometry (VLBI), and satellite laser ranging (SLR).

Answer: (c)

10. Explanation:

Surface fitting is the best approach to datum transformation. The United States, Canada, and Australia use this method. Although the number of common points available is an important element in evaluating the accuracy that can be expected from a datum transformation method, generally the Molodenski transformation provides the least accurate results. The surface-fitting or grid-shift technique, like that used in NADCON, will provide the most accurate transformation.

Answer: **(c)**

3

Heights

Latitude and longitude, northing and easting, radius vector and polar angle, and coordinates often come in pairs—but that is not the whole story. For a coordinate pair to be entirely accurate, the point it represents must lie on a well-defined surface. It might be a flat plane, or it might be the surface of a particular ellipsoid; in either case, the surface will be smooth and have a definite and complete mathematical definition.

As mentioned before, modern geodetic datums rely on the surfaces of geocentric ellipsoids to approximate the surface of the Earth. But the actual Earth does not coincide with these nice smooth surfaces, even though that is where the points represented by the coordinate pairs lie. In other words, the abstract points are on the ellipsoid, but the physical features those coordinates intend to represent are, of course, on the actual Earth. Though the intention is for the Earth and the ellipsoid to have the same center, the surfaces of the two figures are certainly not in the same place. There is a distance between them.

Ellipsoidal Height

The distance represented by a coordinate pair on the reference ellipsoid to the point on the surface of the Earth is measured along a line perpendicular to the ellipsoid. This distance is known by more than one name. It is called the *ellipsoidal height* and is also called the *geodetic height*, and it is usually symbolized by h.

In Figure 3.1, the ellipsoidal height of station Youghall is illustrated. The reference ellipsoid is GRS80, since the latitude and longitude are given in NAD83. The concept of an ellipsoidal height is straightforward. A reference ellipsoid may be above or below the surface of the Earth at a particular place. If the ellipsoid's surface is below the surface of the Earth at the point, the ellipsoidal height has a positive sign; if the ellipsoid's surface is above the surface of the Earth at the point, the ellipsoidal height has a negative sign. It is important to remember that the measurement of an ellipsoidal height is along a line perpendicular to the ellipsoid, not along a plumb line. Said another way, an ellipsoidal height is not measured in the direction of gravity. It is not measured in the conventional sense of down or up.

Ellipsoidal Height = h

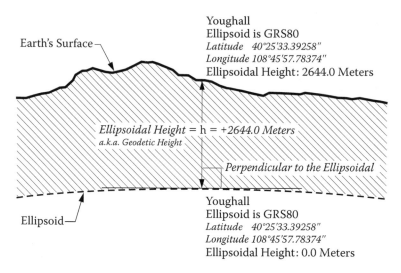

FIGURE 3.1
Ellipsoidal height.

As mentioned in Chapter 1, "down" is a line perpendicular to the ellipsoidal surface at a particular point on the ellipsoidal model of the Earth. On the real Earth, down is the direction of gravity at the point. Most often they are not the same. And since a reference ellipsoid is a geometric imagining, it is quite impossible to actually set up an instrument on it. That makes it tough to measure ellipsoidal height using surveying instruments. In other words, ellipsoidal height is not what most people think of as an elevation.

Nevertheless, the ellipsoidal height of a point is readily determined using a GPS (Global Positioning System) receiver. GPS can be used to discover the distance from the geocenter of the Earth to any point on the Earth, or above it for that matter. In other words, it has the capability of determining three-dimensional coordinates of a point in a short time. It can provide latitude and longitude, and—if the system has the parameters of the reference ellipsoid in its software—it can calculate the ellipsoidal height. The relationship between points can be further expressed in the ECEF coordinates, x, y, and z, or in a *Local Geodetic Horizon System* (*LHGS*) of north, east, and up. Actually, in a manner of speaking, ellipsoidal heights are new, at least in common usage, since they could not be easily determined until GPS became a practical tool in the 1980s. However, ellipsoidal heights are not all the same, because reference ellipsoids and sometimes their origins can differ. For example, an ellipsoidal height expressed in ITRF00 would be based on an ellipsoid with exactly the same shape as the NAD83 ellipsoid, GRS80; nevertheless, the heights would be different because the origin has a different relationship with the Earth's surface (see Figure 3.2).

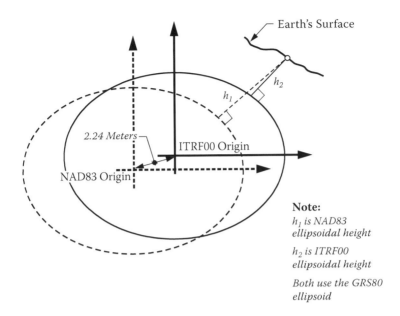

Note:

h_1 is NAD83 ellipsoidal height

h_2 is ITRF00 ellipsoidal height

Both use the GRS80 ellipsoid

FIGURE 3.2
All ellipsoidal heights are not the same.

There is nothing new about heights themselves, or elevations as they are often called. Long before ellipsoidal heights were so conveniently available, knowing the elevation of a point was critical to the complete definition of a position. In fact, there are more than 200 different vertical datums in use in the world today. They were, and still are, determined by a method of measurement known as *leveling*. But it is important to note that this process measures a very different sort of height.

Both *trigonometric leveling* and *spirit leveling* depend on optical instruments. Their lines of sight are oriented to gravity, not a reference ellipsoid. Therefore, the heights established by leveling are not ellipsoidal. In fact, a reference ellipsoid actually cuts across the level surfaces to which these instruments are fixed.

Trigonometric Leveling

Finding differences in heights with trigonometric leveling requires a level optical instrument that is used to measure angles in the vertical plane, a graduated rod, and either a known horizontal distance or a known slope distance between them. As shown in Figure 3.3, the instrument is centered over a point of known elevation, and the rod is held vertically on the point of

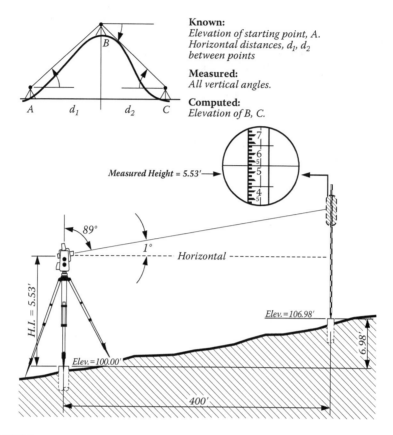

FIGURE 3.3
Trigonometric leveling.

unknown elevation. At the instrument, one of two angles is measured: either the vertical angle, from the horizontal plane of the instrument, or the zenith angle, from the instrument's vertical axis. Either angle will do. This measured angle, together with the distance between the instrument and the rod, provides two known components of a right triangle in the vertical plane. It is then possible to solve that triangle to reveal the vertical distance between the point at the instrument and the point on which the rod is held.

For example, suppose that the height or elevation of the point over which the instrument is centered is 100.00 ft. Further, suppose that the height of the instrument's level line of sight, its horizontal plane, is 5.53 ft above that point. The *height of the instrument* (H.I.) would then be 105.53 ft. For convenience, the vertical angle at the instrument could be measured to 5.53 ft on the rod. If the measured angle is 1° 00′ 00″ and the horizontal distance from the instrument to the rod is known to be 400.00 ft, all the elements are in place to calculate a new height. In this case, the tangent of 1° 00′ 00″ multiplied by 400.00 ft yields 6.98 ft, which is the difference in height from the point at the instrument and the point at the rod. Therefore, 100.00 ft plus 6.98 ft indicates a height of 106.98 at the new station where the rod was placed.

There are many more aspects to this process—the curvature of the Earth, refraction of light, etc.—that make it much more complex in practice than shown in this illustration. However, the fundamental of the procedure is the solution of a right triangle in a vertical plane using trigonometry, hence the name trigonometric leveling. It is faster and more efficient than spirit leveling, but not as precise.

Horizontal surveying usually precedes leveling in control networks. This was true in the early days of what has become our national network, the *National Spatial Reference System* (*NSRS*) of the United States. Geodetic leveling was begun only after triangulation networks were under way. This was also the case in many other countries. In some places around the world, the horizontal work was even completed before leveling was commenced. In the United States, trigonometric leveling was applied to geodetic surveying before spirit leveling. Trigonometric leveling was used extensively to provide elevations to reduce the angle observations and base lines necessary to complete triangulation networks to *sea level*. Moreover, the angular measurements for the trigonometric leveling were frequently done in an independent operation with instruments having only a vertical circle.

Then, in 1871, Congress authorized a change for what was then called the *Coast Survey* under Benjamin Peirce that brought spirit leveling to the forefront. The Coast Survey was to begin a transcontinental arc of triangulation to connect the surveys on the Atlantic coast with those on the Pacific coast. Until that time, their work had been restricted to the coasts. But with the undertaking of triangulation that would cross the continent along the 39th parallel, it was clear that trigonometric leveling was not sufficient to support the project. The survey needed more vertical accuracy than it could provide. So in 1878, at about the time the work began, the name of the agency was changed from Coast Survey to U.S. Coast and Geodetic Survey, and a line of spirit leveling of high precision was begun at *Benchmark A* in the foundation wall of the Washington County Court House in Hagerstown, Maryland. It headed west, and it finally reached the west coast at Seattle in 1907. Along the way, it provided benchmarks for the use of engineers and others who needed accurate elevations or heights for subsequent work, and established the vertical datum for the United States.

Spirit Leveling

The spirit-leveling method (see Figure 3.4) is simple in principle, but not in practice. An instrument called a *level* is used to establish a line of sight that is perpendicular to gravity, in other words, a level line. Then two rods marked with exactly the same graduations, like rulers, are held vertically, resting on two solid points, one ahead and one behind the level along the route of the survey. The system works best when the level is midway between these rods. Looking at the rod to the rear through the telescope of the level, the point at which the horizontal level line of sight of the level intersects the vertical rod, there is a graduation. That reading is taken and noted. This is known as the *backsight* (B.S.). This reading tells the height, or elevation, that the line of sight of the level is above the mark on which the rod is resting. For example, if the point on which the rod is resting is at an elevation of 100 ft and the reading on the rod is 6.78 ft, the height of the level's line of sight is 106.78 ft. That value is known as the H.I. Then the still-level instrument is rotated to observe the vertical rod ahead and a value is read there. This is known as the *foresight* (F.S.). The difference between the two readings reveals the change in elevation from the first point at the backsight to the second at the foresight. For example, if the first reading established the height of the level's line of sight, the H.I., at 106.78 ft, and the reading on the rod ahead, the F.S., was 5.67 ft it becomes clear that the second mark is 1.11 ft higher than the first. It has an elevation of 101.11 ft. By beginning this process from a monumented point of known height, a benchmark, and repeating it with

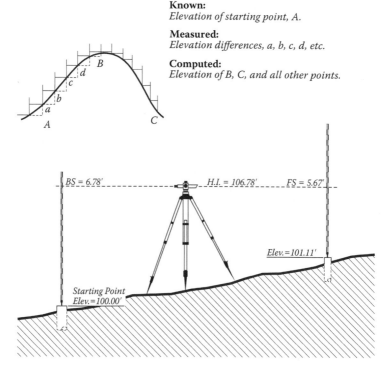

Known:
Elevation of starting point, A.

Measured:
Elevation differences, a, b, c, d, etc.

Computed:
Elevation of B, C, and all other points.

FIGURE 3.4
Spirit leveling.

good procedures, the heights of marks can be determined all along the route of the survey.

The accuracy of level work depends on the techniques and the care used. Methods such as balancing the forward and back sights, calculating refraction errors, running new circuits twice, using one-piece rods, etc., can improve results markedly. In fact, entire books have been written on the details of proper leveling techniques. Here the goal will be to mention only a few elements pertinent to coordinates generally.

It is difficult to overstate the amount of effort devoted to differential spirit-level work that has carried vertical control across the United States. The transcontinental precision leveling surveys done by the Coast and Geodetic Survey from coast to coast were followed by thousands of miles of spirit-leveling work of varying precision. When the 39th parallel survey reached the West Coast in 1907, there were approximately 19,700 miles (31,789 km) of geodetic leveling in the national network. That was more than doubled 22 years later, in 1929, to approximately 46,700 miles (75,159 km). As the quantity of leveling information grew, so did the errors and inconsistencies. The foundation of the work was ultimately intended to be mean sea level (MSL) as measured by *tide station gauges*. Inevitably, this growth in leveling information and benchmarks made a new general adjustment of the network necessary to bring the resulting elevations closer to their true values relative to mean sea level.

There had already been four previous general adjustments to the vertical network across the United States by 1929. They were completed in 1900, 1903, 1907, and 1912. The adjustment in 1900 was based upon elevations held to mean sea level as determined at five tide stations. The adjustments in 1907 and 1912 left the eastern half of the United States fixed as adjusted in 1903. In 1927 there was a special adjustment of the leveling network. This adjustment was not fixed to mean sea level at all tide stations, and after it was completed, it became apparent that the mean sea level surface as defined by tidal observations had a tendency to slope upward to the north along both the Pacific and Atlantic Coasts, with the Pacific higher than the Atlantic.

In the adjustment that established the Sea Level Datum of 1929, the determinations of mean sea level at 26 tide stations—21 in the United States and 5 in Canada—were held fixed. Sea level was the intended foundation of these adjustments, and it might make sense to say a few words about the forces that shape it.

Sea Level

Both the Sun and the Moon exert tidal forces on the Earth, but the Moon's force is greater. The Sun's tidal force is about half of that exerted on the Earth by the Moon. The Moon makes a complete elliptical orbit around the Earth every 27.3 days. There is a gravitational force between the Moon and the Earth. Each pulls on the other, and at any particular moment, the gravitational pull is greatest on the portion of the Earth that happens to be closest to the Moon. That produces a bulge in the waters on the Earth in response to the tidal force. On the side of the Earth opposite the bulge, centrifugal force exceeds the gravitational force of the Earth and water in this area is forced out, away from the surface of the Earth, creating another bulge. But the two bulges are not stationary; they move across the surface of the Earth. They move because not only is the Moon moving slowly relative to the Earth as it proceeds along its orbit, but more importantly, the Earth is rotating in relation to the Moon. And the Earth's rotation is relatively rapid in comparison with the Moon's movement. Therefore, a coastal area in the high middle latitudes may find itself with a high tide early in the day when it is close to the Moon and then a low tide in the middle of the day when it has rotated away from it. This cycle will begin again with another high tide a bit more than 24 hours after the first high tide. The interval is a bit more than 24 hours because the time from the moment the Moon reaches a particular meridian to the next time it is there is actually about 24 hours and 50 min, a period called a *lunar day*.

This sort of tide, with one high water and one low water in a lunar day, is known as a *diurnal tide*. This characteristic tide would be most likely to occur in the middle latitudes to the high latitudes when the Moon is near its maximum declination, as you can see from Figure 3.5.

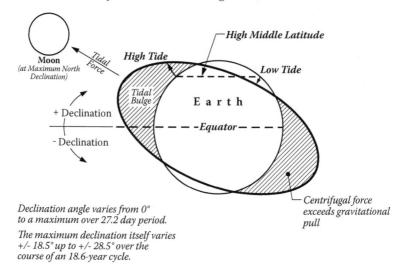

Declination angle varies from 0° to a maximum over 27.2 day period.

The maximum declination itself varies +/- 18.5° up to +/- 28.5° over the course of an 18.6-year cycle.

FIGURE 3.5
Diurnal tide.

The declination of a celestial body is similar to the latitude of a point on the Earth. It is an angle measured at the center of the Earth from the plane of the equator, positive to the north and negative to the south, to the subject, which is in this case the Moon. The Moon's declination varies from its minimum of 0° at the equator to its maximum over a 27.2-day period, and that maximum declination oscillates too. It goes from ±18.5° up to ±28.5° over the course of an 18.6-year cycle.

Another factor that contributes to the behavior of tides is the elliptical nature of the Moon's orbit around the Earth. When the Moon is closest to the Earth, i.e., at *perigee*, the gravitational force between the Earth and the Moon is 20% greater than usual. At *apogee*, when the Moon is farthest from the Earth, the force is 20% less than usual. The variations in the force have exactly the effect you would expect on the tides, making them higher and lower than usual. It is about 27.5 days from perigee to perigee.

To summarize, the Moon's orbital period is 27.3 days. It also takes 27.2 days for the Moon to move from its maximum declinations back to 0° directly over the equator. And there are 27.5 days from one perigee to the next. You can see that these cycles are almost the same—almost, but not quite. They are just different enough that it takes 18 to 19 years for the Moon to go through all the possible combinations of its cycles with respect to the Sun and the Earth. Therefore, if you want to be certain that you have recorded the full range of tidal variation at a place you must observe and record the tides at that location for 19 years.

This 19-year period, sometimes called the *Metonic cycle*, is the foundation of the definition of mean sea level (MSL). Mean sea level can be defined as the arithmetic mean of hourly heights of the sea at a primary-control tide station observed over a period of 19 years. The *mean* in mean sea level refers to the average of these observations over time at one place. It is important to note that it does not refer to an average calculation made from measurements at several different places. Therefore, when the Sea Level Datum of 1929 was fixed to MSL at 26 tide stations, that meant it was made to fit 26 different and distinct local MSLs. In other words, it was warped to coincide with 26 different elevations.

The topography of the sea changes from place to place, and that means, for example, that MSL in Florida is not the same as MSL in California. The fact is, mean sea level varies. And the water's temperature, salinity, currents, density, wind, and other physical forces all cause changes in the sea surface's topography. For example, the Atlantic Ocean north of the Gulf Stream's strong current is around 1 m lower than it is farther south. And the denser water of the Atlantic is generally about 40 cm lower than the Pacific. At the Panama Canal, the actual difference is about 20 cm from the east end to the west end.

Evolution of the Vertical Datum

After it was formally established, thousands of miles of leveling were added to the Sea Level Datum of 1929 (SLD29). The Canadian network also contributed data to the Sea Level Datum of 1929, but Canada did not ultimately use what eventually came to be known as the National Geodetic Vertical Datum of 1929 (NGVD29). The name was changed in 1973 because the final result did not really coincide with mean sea level. It became apparent that the precise leveling done to produce the fundamental data had great internal consistency, but that consistency suffered when the network was warped to fit so many tide station determinations of mean sea level.

By the time the name was changed to NGVD29, there were more than 400,000 miles of new leveling work included. There were distortions in the network. Original benchmarks had been disturbed, destroyed, or lost. The National Geodetic Survey (NGS) thought it time to consider a new adjustment. This time, there was a different approach. Instead of fixing the adjustment to tidal stations, the new adjustment would be minimally constrained. That meant that it would be fixed to only 1 station, not 26. That station was Father Point/Rimouski, an *International Great Lakes Datum of 1985 (IGLD85)* station near the mouth of the St. Lawrence River and on its southern bank. In other words, for all practical purposes, the new adjustment of the huge network was not intended to be a sea level datum at all. It was a change in thinking that was eminently practical.

While is it relatively straightforward to determine mean sea level in coastal areas, carrying that reference reliably to the middle of a continent is quite another matter. Therefore, the new datum would not be subject to the variations in sea surface topography. It was unimportant whether the new adjustment's zero elevation and mean sea level were the same.

Zero Point

At this stage, it is important to mention that, throughout the years, there were, and continue to be, benchmarks set and vertical control work done by official entities in federal, state, and local governments other than NGS. State departments of transportation, city and county engineering and public works departments, the U.S. Army Corps of Engineers, and many other governmental and quasi-governmental organizations have established their own vertical control networks. Included on this list is the U.S. Geological Survey (USGS). In fact, minimizing the effect on the widely used USGS mapping products was an important consideration in designing the new datum adjustment. Several of these agencies, including the *National Oceanic and*

Atmospheric Administration (*NOAA*), the U.S. Army Corps of Engineers, the Canadian Hydrographic Service, and the Geodetic Survey of Canada worked together for the development of the International Great Lakes Datum 1985 (IGLD85). This datum was originally established in 1955 to monitor the level of the water in the Great Lakes.

Precise leveling proceeded from the zero reference established at Pointe-au-Père, Quebec, in 1953. The resulting benchmark elevations were originally published in September 1961. The result of this effort was the International Great Lakes Datum 1955. After nearly 30 years, the work was revised. The revision effort began in 1976, and the result was IGLD85. It was motivated by several developments, including deterioration of the zero-reference-point gauge location and improved surveying methods. But one of the major reasons for the revision was the movement of previously established benchmarks due to *isostatic rebound*. This effect is, literally, the Earth's crust rising slowly, rebounding, from the removal of the weight and subsurface fluids caused by the retreat of the glaciers from the last ice age.

The choice of the tide gauge at Pointe-au-Père, Quebec, as the zero reference for IGLD was logical in 1955. It was reliable, it had already been connected to the network with precise leveling, and it was at the outlet of the Great Lakes. But by 1984, the wharf at Pointe-au-Père had deteriorated, and the gauge was eventually moved about 3 miles to Rimouski, Quebec, and precise levels were run between the two. It was there that the zero reference for IGLD85—and what then became a new adjustment called North American Vertical Datum 1988 (NAVD88)—was established.

The readjustment, known as NAVD88, was begun in the 1970s. It addressed the elevations of benchmarks all across the nation. The effort also included field work. Destroyed and disturbed benchmarks were replaced. The miles of leveling data increased from 46,701 (75,159 km) used in the establishment of NGVD29 to 622,303 miles (1,001,500 km) used in the establishment of NAVD88, which was ready in June of 1991. The differences between elevations of benchmarks determined in NGVD29 compared with the elevations of the same benchmarks in NAVD88 vary from approximately –1.3 ft in the east to approximately +4.9 ft in the west in the 48 conterminous United States. The larger differences tend to be on the coasts, as one would expect, since NGVD29 was forced to fit mean sea level at many tidal stations, while NAVD88 was held only to one.

When comparing heights in IGLD85 and NAVD88, it is important to consider that they are both based on the zero point at Father Point/Rimouski. There is really only one difference between the nature of the heights in the two systems. NAVD88 values are expressed in *Helmert orthometric height* units, and IGLD85 elevations are given in *dynamic height* units. The explanation of this difference requires introduction of some important principles of the current understanding of heights.

So far, there has been mention of heights based on the ellipsoidal model of the Earth and heights that use mean sea level as their foundation. While ellipsoidal heights are not affected by any physical forces at all, heights based on mean sea level are affected by a broad range of them. There is also another surface to which heights are referenced that is defined by only one force, gravity. It is known as the *geoid*.

Geoid

Any object in the Earth's gravitational field has *potential energy* derived from being pulled toward the Earth. Quantifying this potential energy is one way to talk about height, because the amount of potential energy an object derives from the force of gravity is related to its height.

Here is another way of saying the same thing. The potential energy an object derives from gravity equals the work required to lift it to its current height. Imagine several objects, each with the same weight, resting on a truly level floor. In that instance, they would all be possessed of the same potential energy from gravity. Their potential energies would be equal. The floor on which they were resting could be said to be a surface of equal potential, an *equipotential* surface.

Now suppose that each of the objects was lifted up onto a level table. It is worth mentioning that they would be lifted through a large number of equipotential surfaces between the floor and the table top, and those surfaces are not parallel with each other. In any case, their potential energies would obviously be increased in the process. Once they were all resting on the table, their potential energies would again be equal, now on a higher equipotential surface—but how much higher?

There is more than one way to answer that question. One way is to find the difference in their *geopotential,* which is their potential energy on the floor, thanks to gravity, compared with their geopotential on the table. This is the same idea behind answering the question with a dynamic height. Another way to answer the question is to simply measure the distance along a plumb line from the floor to the tabletop. This latter method is the basic idea behind an orthometric height. An orthometric height can be illustrated by imagining that the floor in the example is a portion of one particular equipotential surface called the geoid.

The geoid is a unique equipotential surface that best fits mean sea level. As discussed previously, mean sea level is not a surface on which the geopotential is always the same; so it is not an equipotential surface at all. Forces other than gravity affect it, forces such as temperature, salinity, currents, wind, etc. On the other hand, the geoid by definition is an equipotential surface. It is defined by gravity alone. Further, it is the particular equipotential surface arranged to fit mean sea level as well as possible, in

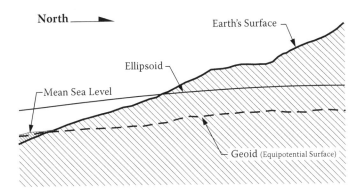

FIGURE 3.6
Mean sea level.

a least squares sense. Across the geoid, the potential of gravity is always the same.

So while there is a relationship between mean sea level and the geoid, they are not the same (see Figure 3.6). They could be the same if the oceans of the world could be utterly still, completely free of currents, tides, friction, variations in temperature, and all other physical forces, except gravity. Reacting to gravity alone, these unattainable calm waters would coincide with the geoid. If the water was then directed by small frictionless channels or tubes and allowed to migrate across the land, the water would then, theoretically, define the same geoidal surface across the continents, too. Of course, the 70% of the Earth covered by oceans is not so cooperative, and the physical forces cannot really be eliminated. These unavoidable forces actually cause mean sea level to deviate up to 1 and 2 m from the geoid.

Because the geoid is completely defined by gravity, it is not smooth. As shown in the exaggerated illustration in Figure 3.7, it is lumpy because gravity is not consistent across the surface of the Earth. It undulates with the uneven distribution of the mass of the Earth. It has all the irregularity that the attendant variation in gravity implies. In fact, the separation between the lumpy surface of the geoid and the smooth GRS80 ellipsoid worldwide varies from about +85 m west of Ireland to about −106 m in the area south of India near Sri Lanka.

At every point, gravity has a magnitude and a direction. Anywhere on the Earth, a vector can describe gravity, but these vectors do not all have the *same* direction or magnitude. Some parts of the Earth are denser than others. Where the Earth is denser, there is more gravity, and the fact that the Earth is not a sphere also affects gravity. It follows then that defining the geoid precisely involves actually measuring the direction and magnitude of gravity at many places—but how?

FIGURE 3.7
Exaggerated representation of the geoid.

Measuring Gravity

Some gravity measurements are executed with a class of instruments called *gravimeters*, which were introduced in the middle of the last century. One sort of gravimeter can be used to measure the relative difference in the force of gravity from place to place. The basic idea of this kind of gravimeter is illustrated by considering the case of one of those weights, described in the earlier analogy, suspended at the end of a spring. Suppose the extension of the spring was carefully measured with the gravimeter on the table and measured again when it was on the floor. If the measurement of such a tiny difference were possible, the spring would be found to be infinitesimally longer on the floor because the magnitude of gravity increases as you move lower. In practice, such a measurement is quite difficult, and thus the increases in the tension on the spring necessary to bring the weight back to a predefined zero point are actually measured.

Conversely, suppose the tension of the spring was carefully measured with the gravimeter on the floor and measured again on the table. The spring would be shortened and the tension on the spring would need to decrease to bring the weight back to the zero point. This is because the spinning of the Earth on its axis creates a *centrifugal*, center-fleeing, force.

Centrifugal force opposes the downward gravitational attraction. And these two forces are indistinguishably bound together. Therefore, measurements made by gravimeters on the Earth inevitably contain both centrifugal and gravitational forces. It is impossible to pry them apart. An idea called the *equivalence principle* states that the effects of being accelerated to a velocity are indistinguishable from the effects of being in a gravitational field. In other words, there is no physical difference between an accelerating frame of reference and the same frame of reference in a gravitational field. So they are taken together. In any case, as you go higher, the centrifugal force increases and counteracts the gravitational attraction to a greater degree than it does at lower heights and so, generally speaking, gravity decreases the higher you go.

Now please recall that the Earth closely resembles an oblate spheroid. That means that the distance from the center of the planet to a point on the equator is longer than the distance from the center to the poles. Said another way, the Earth is generally higher at the equator than it is at the poles. As a consequence, the acceleration of a falling object is less at the equator than at the poles. Less *acceleration of gravity* means that if you drop a ball at the equator, the rate at which its fall would accelerate would be less than if you dropped it at one of the poles. As a matter of fact, that describes the basic idea behind another kind of gravimeter. In this second kind of gravimeter, the fall of an object inside a vacuum chamber is very carefully measured.

The acceleration of gravity, i.e., the rate at which a falling object changes its velocity, is usually quantified in *gals,* a unit of measurement named for Galileo, who pioneered the modern understanding of gravity. What is a gal? Well, imagine an object is dropped. At the end of the first second, it is falling at 1 cm/sec. Then at the end of the next second it is falling at 2 cm/sec. In this thought experiment, the imagined object would have accelerated 1 gal. Said another way, 1 gal is an acceleration of 1 cm/sec/sec.

At the equator, the average acceleration of a falling object is approximately 978 gals, which is 978 cm/sec^2 or 32.09 ft/sec^2. At the poles, the acceleration of a falling object increases to approximately 984 gals, which is 984 cm/sec^2 or 32.28 ft/sec^2. The acceleration of a falling object at 45° latitude is between these two values, as you would expect. It is 980.6199 gals. This value is sometimes called *normal gravity.*

Orthometric Correction

This increase in the rate of acceleration due to gravity between the equator and the poles is a consequence of the increase in the force of gravity as the Earth's surface gets closer to the center of mass of the Earth. Imagine the equipotential surfaces that surround the center of mass of the Earth as the layers of an onion. These layers are farther apart at the equator than they are at the poles. This is because there is a larger centrifugal force at 0° latitude compared with the centrifugal force at 90° latitude. In other words, equipotential surfaces get closer together as you approach the poles, i.e., they converge, as seen in Figure 3.8. And the effect of this convergence becomes more pronounced as the direction of your route gets closer to north and south.

This effect was mentioned as far back as 1899. It was discovered that the precise leveling run in an east-west direction required less correction than leveling done in a north or south direction. Eventually, a value known as an *orthometric correction* was applied to accommodate the convergence of the equipotential surfaces. As Howard Rappleye wrote,

> The instruments and methods used in 1878 were continued in use until 1899, when, as the result of an elaborate theoretical investigation, it was found that apparently the leveling was subject to a systematic error depending on the azimuth of the line of levels.... The leveling of the Coast and Geodetic Survey was then corrected for the systematic error. (Rappleye 1948, 1)

In the years that led to the establishment of NGVD29, gravity data was mostly unavailable, and the actual correction applied was based on an ellipsoidal model. The result is known as *normal orthometric heights,* which do not take account of local variations in gravity.

The application of the *orthometric correction* means that the height difference derived from leveling between two points will not exactly match the difference between orthometric heights. This is a consequence of the fact that the line of sight of a properly balanced level will follow an equipotential surface, i.e., a level surface, but orthometric heights after their correction are not exactly on a level surface. Consider two points, two benchmarks, with

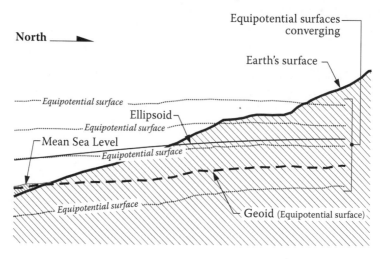

FIGURE 3.8
Equipotential surfaces converging.

the same published orthometric height, one north and one south. The orthometric height of the south benchmark as measured along a plumb line from the geoid will pass through fewer level surfaces than the same measurement to the benchmark at the north end. For example, the orthometric height of the water surface at the south end of Lake Huron seems to indicate that it is approximately 5 cm higher than the same equipotential surface at the north end. The orthometric heights make it look this way because the equipotential surfaces are closer together at the north end than they are at the south end. But precise levels run from the south end of the lake to the north would not reflect the 5-cm difference because the line of sight of the level would actually follow the equipotential surface all the way.

In other words, the convergence of equipotential surfaces prevents leveling from providing the differences between points as it is defined in orthometric heights. And the amount of the effect depends on the direction of the level circuit. The problem can be alleviated somewhat by applying an orthometric correction based on the measurement of gravity. An orthometric correction can amount to 0.04 ft/mile in mountainous areas. It is systematic and is not eliminated by careful leveling procedures.

It is worthwhile to note that NGS publishes height data as Helmert orthometric heights. This is a particular type of orthometric height that does not take account of the gravitational effect of topographic relief. As a consequence, these heights can lead to a certain level of misclosure between GPS-determined benchmarks and the geoid model in mountainous areas.

Ellipsoidal, Geoidal, and Orthometric Heights

The distance measured along a line perpendicular to the ellipsoid from the ellipsoid of reference to the geoid is known as a *geoid height*. It is usually symbolized as N. In the conterminous United States, sometimes abbreviated CONUS, geoid heights vary from about −8 m to about −53 m in NAD83. These are larger than those in the old NAD27 system. Please recall that its orientation at Meades Ranch, Kansas, was arranged so that the distance between the Clarke 1866 ellipsoid and the geoid was zero. And across the United States, the difference between them in NAD27 never grew to more than 12 m. In fact, for all practical purposes, the ellipsoid and the geoid were often assumed to coincide in that system. However, in NAD83, based on the GRS80 ellipsoid, the geoid heights are larger and negative. If the geoid is above the ellipsoid, N is positive; if the geoid is below the ellipsoid, N is negative. Throughout the conterminous United States, the geoid is underneath the ellipsoid. In Alaska, it is the other way around, and the ellipsoid is underneath the geoid, and N is positive.

As shown in Figure 3.9, ellipsoidal height is symbolized as *h*. The ellipsoidal height is also measured along a line perpendicular to the ellipsoid of reference, but to a point on the surface of the Earth. An orthometric height, symbolized, *H*, is measured along a plumb line from the geoid to a point on the surface of the Earth.

In either case, by using the formula,

$$H = h - N$$

one can convert an ellipsoidal height, *h*, derived, say, from a GPS observation, into an orthometric height, *H*, by knowing the extent of geoid–ellipsoid separation, i.e., the geoid height, *N*, at that point.

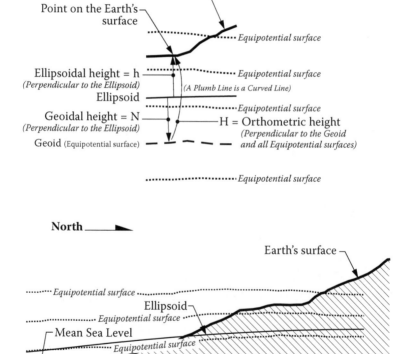

FIGURE 3.9
Orthometric height (*H*), ellipsoidal height (*h*), geoidal height (*N*).

As you can see from Figure 3.9, the ellipsoidal height of a particular point is actually smaller than the orthometric height throughout the conterminous United States.

The formula $H = h - N$ does not account for the fact that the plumb line along which an orthometric height is measured is curved, as seen in Figure 3.9. It is curved because it is perpendicular to every equipotential surface through which it passes. And since equipotential surfaces are not parallel with each other, the plumb line must be a curve to maintain perpendicularity with them. This deviation of a plumb line from the perpendicular to the ellipsoid reaches about 1 min of arc in only the most extreme cases. Therefore, any height difference that is caused by the curvature is negligible. It would take a height of over 6 miles for the curvature to amount to even 1 mm of difference in height.

It might be pertinent to ask, "Why use orthometric heights at all?" One answer is the accommodation of GPS measurements. Orthometric elevations are not directly available from the geocentric position vectors derived from GPS measurements. However, they can be rather quickly calculated using $H = h - N$. That is, they can be calculated once a geoidal model is well defined.

NGS GEOID Models

The geoid defies the certain, clear definition of, say, the GRS80 ellipsoid. It does not precisely follow mean sea level, and neither does it exactly correspond with the topography of the dry land. It is irregular like the terrestrial surface, and has similar peaks and valleys, but they are due to the uneven distribution of the mass of the planet. The undulations of the geoid are defined by gravity and reflect changes in density known as *gravity anomalies*. Gravity anomalies represent the difference between the equipotential surface of so-called *normal gravity*, which theoretically varies with latitude, and actual measured gravity at a place. The calculation of geoid heights using gravity anomalies, Δg, is usually done with the formula derived in 1849 by George Stokes, but its effectiveness depends on the accuracy of the modeling of the geoid around the world.

With the major improvements in the mapping of the geoid on both national and global scales over the past quarter century, geoid modeling has become more and more refined. To some degree, this is due to improvements in data gathering. For example, gravimeter surveys on land routinely detect gravity anomalies to a precision of 1 part in 1 million. Surveys with a precision of 0.01 milligal, that is one hundredth of one thousandth of a gal, are common. Moreover, GPS allows the accurate positioning of gravimetric stations. Further, there is the advantage that there is good deal of gravimetric information available through governmental agencies and universities around the world, though the distribution of the data may not be optimal. And new satellite altimetry missions also contribute to the refinement of geoid modeling.

New and improved data sources have led to applications built on better computerized modeling at NGS. For many years, these applications have been allowing users to easily calculate values for N, the geoid height, which is the distance between the geoid and the ellipsoid, at any place as long as its latitude and longitude are available. The computer model of the geoid has been steadily improving. The latest of these to become available is known as GEOID03.

The GEOID90 model was rolled out at the end of 1990. It was built using over a million gravity observations. The GEOID93 model, released at the beginning of 1993, utilized many more gravity values. Both provided geoid heights in a grid of 3 min of latitude by 3 min of longitude, and their accuracy was about 10 cm. Next, the GEOID96 model, with a grid of 2 min of latitude by 2 min of longitude, was released. More recently GEOID99 was available to cover the conterminous United States, and includes the U.S. Virgin Islands, Puerto Rico, Hawaii, and Alaska. It was computed using 2.6 million gravity measurements. The grid was 1 min of latitude by 1 min of longitude. In practical terms, GEOID96 was matched to NAVD88 heights on about 3,000 benchmarks with an accuracy of about ±5.5 cm (1 sigma), whereas GEOID99 was matched to NAVD88 heights on about 6,000 benchmarks with an accuracy of about ±4.6 cm (1 sigma). GEOID99 was the first of the models to combine gravity values with GPS ellipsoidal heights on previously leveled benchmarks. Therefore, users relying on GEOID99 could trust its representation of the relationship between GPS ellipsoid heights in NAD83 with orthometric heights in the NAVD88 datum.

GEOID03 is also a model of the conterminous United States (CONUS). It supersedes the previously mentioned models. It was built with a combination of gravity data and ellipsoid heights derived from GPS at 14,185 leveled benchmarks, including 579 in Canada. In Alaska, there was a shortage of such information. Generally, GEOID03 can provide data valid to about ±2.4 cm (1 sigma) for the conversion between NAD83 GPS ellipsoidal heights and NAVD88 orthometric heights. The state with the smallest standard deviation in this regard is Connecticut (1.3 cm), and that with the largest is Texas (5.8 cm). Nationwide, GEOID03 is a 50% improvement over GEOID99. This improvement is due, in part, to the use of a more complicated analytic function in the development of GEOID03 than was available for GEOID99. Differences of 10–15 cm are possible in some coastal and mountainous areas between the two models. It is always good practice to include existing benchmarks in GPS surveys so that the difference between their published elevations and the heights derived through the use of the GEOID03 model can be compared. This is especially true in areas such as the western states, where the sparseness of data restricted the ability to refine the model, and in areas where subsidence is significant.

Dynamic Heights

Now, please recall that an orthometric height is a measurement along a plumb line from a particular equipotential surface to a point on the Earth's surface. In other words, the orthometric height of that point is its distance from the reference surface, a distance measured along the line perpendicular to every equipotential surface in between. And recall that these equipotential surfaces are not parallel to each other, chiefly because of gravity anomalies and the rotation and shape of the Earth. Therefore, it follows that two points could actually have the same orthometric height and not be on the same equipotential surface. This is a rather odd fact, and it has an unusual implication. It means that water might actually flow between two points that have exactly the same orthometric height.

This is one reason that the International Great Lakes Datum of 1985 is based on *dynamic heights*. Unlike orthometric heights, two points with identical dynamic heights are definitely on the same equipotential surface. Two points would have to have different dynamic heights for water to flow between them, and the flow of water is a critical concern for those using that system.

Because one can rely on the fact that points with the same dynamic heights are always on the same equipotential surface, they are better indicators of the behavior of water than orthometric heights.

An advantage of using dynamic heights is the sure indication of whether a water surface, or any other surface for that matter, is truly level or not. For example, the Great Lakes are monitored by tide gauges to track historical and predict future water levels in the lakes, and it is no surprise then that the subsequent International Great Lakes Datum of 1985 (IGLD85) heights are expressed as dynamic heights. Where the lake surfaces are level, the dynamic heights are the same, and where they are not, it is immediately apparent because the dynamic heights differ.

Points on the same equipotential surface also have the same *geopotential numbers* along with the same dynamic heights. The idea of measuring geopotential by using geopotential numbers was adopted by the International Association of Geodesy in 1955. The geopotential number of a point is the difference between the geopotential below the point, down on the geoid, and the geopotential right at the point itself. Said another way, the geopotential number expresses the work that would be done if a weight were lifted from the geoid up to the point, like the weights that were lifted onto a table in the earlier analogy. A geopotential number is expressed in *geopotential units*, or *gpu*. A gpu is 1 kilogal/m.

Geopotential numbers, along with a constant, contribute to the calculation of dynamic heights. The calculation itself is simple:

$$H\frac{dyn}{p} = \frac{C_p}{\gamma_0}$$

where $H_p{}^{dyn}$ is the dynamic height of a point in meters, C_p is the geopotential number at that point in gpu (i.e., kilogals per meter), and γ_0 is the constant 0.9806199 kgals. The constant is normal gravity at 45° latitude on GRS80.

The dynamic height of a point is found by dividing its NAVD88 geopotential number by the normal gravity value. In other words, dynamic heights are geopotential numbers scaled by a particular constant value chosen in 1984 to be normal gravity at 45° latitude on the GRS80 reference ellipsoid. The whole point of the calculation is to transform the geopotential number that is in kilogals per meter into a dynamic height in meters by dividing by the constant in kilogals. Here is a sample calculation of the dynamic height of station M 393, an NGS benchmark.

$$H^{dyn} = \frac{C}{\gamma_0}$$

$$H^{dyn} = \frac{1660.419936 gpu}{0.9806199 kgals}$$

$$H^{dyn} = 1693.235 \text{ m}$$

The NAVD88 orthometric height of this benchmark determined by spirit leveling is 1694.931 m and differs from its calculated dynamic height by 1.696 m. However, using the same formula and the same geopotential number, but divided by a gravity value derived from the Helmert height reduction formula, the result would be a Helmert orthometric height for M 393, instead of its dynamic height. Note that the geopotential number stays the same in both systems. And there is a third result: If the divisor were a gravity value calculated with the international formula for normal gravity, the answer would be the normal orthometric height for the point.

Reference

Rappleye, H. S. 1948. *Manual of geodetic leveling*. Special Publication 239. Washington, DC: U.S. Department of Commerce, Coast and Geodetic Survey.

Exercises

1. Which of the following can be depended upon to define the flow of water?

   a. From a higher ellipsoidal height to a lower ellipsoidal height

   b. From a higher geodetic height to a lower geodetic height

   c. From a higher orthometric height to a lower orthometric height

   d. From a higher dynamic height to a lower dynamic height

2. Which of the following formulae correctly represents the basic relationship between geodetic, geoidal, and orthometric heights?

   a. $H = h + N$

   b. $h = H + N$

   c. $N = H + h$

   d. None of the above

3. Until the 1940s, the Coast and Geodetic Survey stamped the elevation of a point on its monuments. In what sense would those elevations be obsolete now?

   a. They would be expressed in feet, not in meters.

   b. The elevations would be based on the Clarke reference ellipsoid.

   c. The elevations would not be in the North American Datum of 1988.

   d. The elevations would be dynamic rather than orthometric.

4. Which of the following heights is not measured along a line that is perpendicular to the ellipsoid?

   a. An ellipsoidal height

   b. A geoid height

   c. A geodetic height

   d. An orthometric height

5. Which of these values would not have been affected by gravity?

   a. An ellipsoidal height of 2729.463 m

   b. 983.124 gal from a gravimeter measurement

   c. The measurement of hourly heights of the sea at a primary-control tide station

    d. An orthometric elevation of 5176.00 ft derived from spirit leveling

6. Which of the following events contributed to a substantial increase in the utilization of spirit leveling over trigonometric leveling in precise geodetic surveying in the United States?

    a. The beginning of triangulation to cross the continent along the 39th parallel

    b. The deterioration of the dock at Pointe-au-Père and the subsequent moving of the IGLD zero point to Rimouski, Quebec

    c. The adjustment that established the Sea Level Datum of 1929

    d. The establishment of GPS as a fully operational system

7. Which of the following does *not* correctly describe an improvement of GEOID03 over its predecessors?

    a. It provides the basis for a more accurate conversion between NAD83 GPS ellipsoidal heights and NAVD88 heights than its predecessors.

    b. It is the result of the same analytical function used to produce GEOID99.

    c. It was matched with NAVD88 heights on 14,185 benchmarks.

    d. It was built using gravity values along with GPS ellipsoidal heights on previously leveled benchmarks.

8. When the heights of benchmarks from a spirit-level circuit are compared with the orthometric heights of the same benchmarks calculated from GPS observations, they may differ even though both are based on NAVD88. Which of the following could *not* contribute to the difference?

    a. The undulation of the GRS80 ellipsoid

    b. Lack of balance between the foresights and the backsights when the spirit level circuit was run

    c. Error in the geoid height calculated from GEOID99

    d. The lack of parallelism in equipotential surfaces

9. Which of the following statements is correct?

    a. In the conterminous United States, the GRS80 ellipsoid is always below the geoid that best fits mean sea level per least squares.

    b. The heights of the Atlantic Ocean and the Pacific Ocean are the same.

c. Mean sea level and the geoid that best fits mean sea level from a least-squares point of view nevertheless deviate from mean sea level up to 2 m at some places.

d. The geoid that best fits mean sea level from a least-squares point of view is always below the GRS80 ellipsoid all around the world.

10. What is a gal?

a. A center fleeing force

b. An equipotential surface of the Earth's gravity field

c. A unit used to measure speed

d. A unit of acceleration equal to 1 cm/sec/sec

Explanations and Answers

1. Explanation:

An interesting point: Water would not necessarily flow from a higher ellipsoidal height to a lower ellipsoidal height. You need to use an equipotential surface for that. Water would definitely flow downhill from a higher dynamic height to a lower dynamic height.

Answer: **(d)**

2. Explanation:

Where N represents the geoidal height between the geoid and the ellipsoid, it is a distance that is measured to the geoid along a line perpendicular to the ellipsoid. If the geoid is above the ellipsoid, N is positive, and if the geoid is below the ellipsoid, N is negative. H represents the orthometric height, which is the positive, upward distance from the geoid to the point on the Earth's surface measured along a plumb line. The sum of these two yields the ellipsoidal height, also known as the geodetic height, h, that is the distance from the reference ellipsoid to the point on the surface of the Earth. If the ellipsoid's surface is below the surface of the Earth at the point, the ellipsoidal height has a positive sign; if the ellipsoid's surface is above the surface of the Earth at the point, the ellipsoidal height has a negative sign.

Answer: **(b)**

3. Explanation:

The disks monumenting stations that had elevations stamped on them before the establishment of NAVD88 have been superseded, and those marked elevations are no longer valid. The monument itself has not moved, but its official elevation has changed nonetheless.

Answer: **(c)**

4. Explanation:

A geodetic height is another name for an ellipsoidal height. Both the ellipsoidal, or geodetic, height and the geoid height are measured along lines perpendicular to the reference ellipsoid. The orthometric height, on the other hand, is not. It is measured along a plumb line, which is perpendicular to all of the equipotential surfaces through which it passes from the geoid to the point on the Earth's surface.

Answer: **(d)**

5. Explanation:

An ellipsoidal height does not have a relationship with gravity, as does an orthometric elevation derived from spirit leveling. The orthometric elevation is measured from the geoid, which is an equipotential surface defined by gravity, and the leveling instrument itself is oriented to gravity when it is properly set up. The variation in tides is certainly affected by the gravity of the Earth, the Moon, and the Sun, as are the measurements of a gravimeter, which are of the acceleration of gravity itself.

Answer: **(a)**

6. Explanation:

Congress mandated that the Coast Survey was to begin a transcontinental arc of triangulation to connect the surveys on the Atlantic coast with those on the Pacific coast in 1871. Trigonometric leveling was not sufficient to support the project, and more vertical accuracy was needed. So the Coast Survey under Benjamin Peirce depended on spirit leveling. Until that time the work had been restricted to the coasts. The work began at the Chesapeake Bay in 1878.

Answer: **(a)**

7. Explanation:

GEOID03 is a model of the conterminous United States (CONUS). It supersedes its predecessors. It was built with a combination of gravity data and ellipsoid heights derived from GPS at 14,185 leveled benchmarks, including 579 in Canada. Generally, GEOID03 can provide data valid to about ±2.4 cm (1 sigma) for the conversion between NAD83 GPS ellipsoidal heights and NAVD88 heights. Nationwide, GEOID03 is a 50% improvement over GEOID99. This improvement is due, in part, to the use of a more complicated analytic function in the development of GEOID03 than was available for GEOID99.

Answer: **(b)**

8. Explanation:

The accuracy of level work depends on the techniques and the care used. Methods such as balancing the foresights and backsights, using one-piece rods, etc., can improve results markedly, and not using these methods can lead to systematic error creeping into the work.

Also, equipotential surfaces are not parallel due to variation in the density of the Earth's crust, the Earth's rotation, and other effects.

The convergence of equipotential surfaces prevents leveling from providing the differences between points as defined in orthometric heights. The amount of this effect depends somewhat on the direction of the level circuit. For example, it is generally more pronounced in a north–south direction.

GEOID99 is a great improvement in geoidal modeling, but it is not perfect. It was matched to NAVD88 heights on about 6,000 benchmarks with an accuracy of about ±4.6 cm. In other words, there are inaccuracies in geoid heights calculated using GEOID99. However, the GRS80, unlike the geoid, does not undulate. It is a smooth mathematical surface.

Answer: **(a)**

9. Explanation:

The separation between the lumpy surface of the geoid and the smooth GRS80 ellipsoid worldwide varies from about +85 m west of Ireland to about −106 m in the area south of India near Sri Lanka. The geoid is sometimes above and sometimes below the GRS80 ellipsoid.

In the conterminous United States, sometimes abbreviated as CONUS, the distances between the geoid and the ellipsoid are less. They vary from about −8 m to about −53 m. The geoid heights are negative and are usually symbolized, N. If the geoid is above the ellipsoid, N is positive; if the geoid is below the ellipsoid, N is negative. It is negative because the geoid is actually underneath the ellipsoid throughout the conterminous United States.

Around the world, mean sea level deviates as much as 1–2 m from the geoid.

Answer: **(c)**

10. Explanation:

The acceleration of gravity—the rate at which a falling object changes its velocity—is usually quantified in *gals*, a unit of measurement named for Galileo, who pioneered the modern understanding of gravity. What is a gal? Well, imagine an object traveling at 1 cm/sec at the end of 1 sec. If that object's speed increased to 2 cm/sec by the end of the next second, it would have accelerated 1 gal. Said another way, 1 gal is an acceleration of 1 cm/sec/sec.

Answer: **(d)**

4

Two Coordinate Systems

State Plane Coordinates

State plane coordinates rely on an imaginary flat reference surface with Cartesian axes. They describe measured positions by ordered pairs, expressed in northings and eastings, or x- and y-coordinates. Despite the fact that the assumption of a flat Earth is fundamentally wrong, calculation of areas, angles, and lengths using latitude and longitude can be complicated, so plane coordinates persist. Therefore, the projection of points from the Earth's surface onto a reference ellipsoid—and finally onto flat maps—is still viable.

In fact, many agencies of government, particularly those that administer state, county, and municipal databases, prefer coordinates in their particular *State Plane Coordinate Systems* (*SPCS*). The systems are, as the name implies, state-specific. In many states, the system is officially sanctioned by legislation. Generally speaking, such legislation allows surveyors to use state plane coordinates to legally describe property corners. It is convenient. A Cartesian coordinate and the name of the officially sanctioned system are sufficient to uniquely describe a position. The same fundamental benefit makes the SPCS attractive to government; it allows agencies to assign unique coordinates based on a common, consistent system throughout their jurisdiction.

Map Projection

State plane coordinate systems are built on *map projections*, a term that refers to representing a portion of the actual Earth on a plane. Used for hundreds of years to create paper maps, use of this system continues, but map projection today is most often a mathematical procedure performed by a computer. However, even in an electronic world, projection cannot be done without distortion.

The problem is often illustrated by trying to flatten part of an orange peel that represents the surface of the Earth. A small part, say, a square a quarter of an inch on the side, can be pushed flat without much noticeable deformation. But when the portion gets larger, problems appear. Suppose that a third of the orange peel is involved: As as the center is pushed down, the edges tear and stretch, or both. And as the peel gets even bigger the tearing becomes more severe. So if a map is drawn on the orange before it is peeled, the map is distorted in unpredictable ways when it is flattened, and it becomes difficult to relate a point on one torn piece with a point on another in any meaningful way.

These are the problems that a map projection needs to solve to be useful. The first problem is that the surface of an ellipsoid, like the orange peel, is *nondevelopable.* In other words, flattening it inevitably leads to distortion that is very difficult to model consistently. So, a useful map projection should start with a surface that is *developable,* a surface that can be flattened without all that unpredictable deformation. It happens that both a paper cone and a cylinder illustrate this idea nicely. However, they are illustrations only—models for thinking about the issues involved. Figure 4.1 shows that if a right circular cone is cut from the bottom to the top through an element that is perpendicular to the base, the cone can then be made completely flat without trouble. The same can be said of a cylinder cut perpendicularly from base to base.

Or one could use the simplest case, a surface that is already developed: a flat piece of paper. If the center of a flat plane is brought tangent to the Earth, a portion of the planet can be mapped on it. In other words, a portion of the Earth can be projected directly onto the flat plane. In fact, this is the typical method for establishing an independent *local coordinate system.* These simple Cartesian systems are convenient and satisfy the needs of small projects. The method of projection, onto a simple flat plane, is based on the idea that a small section of the Earth, as with a small section of the orange mentioned previously, conforms so nearly to a plane that distortion on such a system is negligible.

Subsequently, local tangent planes have long been used by land surveyors. Such systems demand little if any manipulation of the field observations, and the approach has merit as long as the extent of the work is small. But the larger the plane grows, the more untenable it becomes. As the area being mapped grows, the reduction of survey observations becomes more complicated, since it must take account of the actual shape of the Earth. This usually involves the ellipsoid, the geoid, and geographical coordinates such as latitude and longitude. At that point, surveyors and engineers rely on map projections to mitigate the situation and limit the now-troublesome distortion. However, a well-designed map projection can offer the convenience of working in plane Cartesian coordinates and still keep distortion at manageable levels.

The design of such a projection must accommodate some awkward facts. For example, while it would be possible to imagine mapping a considerable portion of the Earth using a large number of small individual planes, like facets of a gem, it is seldom done because, when these planes are brought together, they cannot be edge-matched accurately (see Figure 4.2). They cannot be joined properly along their borders. This problem is unavoidable because the planes, tangent at their centers, inevitably depart more and

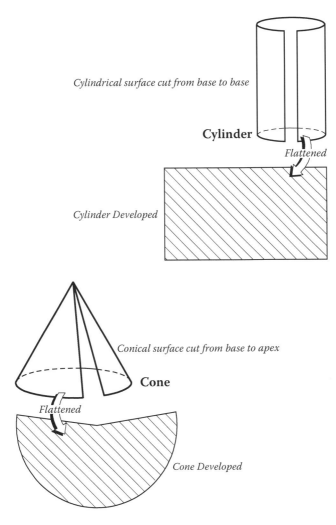

Cylindrical surface cut from base to base

Cylinder

Flattened

Cylinder Developed

Conical surface cut from base to apex

Cone

Flattened

Cone Developed

FIGURE 4.1
The development of a cylinder and a cone.

more from the reference ellipsoid at their edges. The greater the distance between the ellipsoidal surface and the surface of the map on which it is represented, the greater is the distortion on the resulting flat map. This is true of all methods of map projection. Therefore, one is faced with the daunting task of joining together a mosaic of individual maps along their edges, where the accuracy of the representation is at its worst. And even if one could overcome the problem by making the distortion the same on two adjoining maps, another difficulty would remain. Typically, each of these planes has a unique coordinate system. The orientation of the axes—the scale and the rotation of each one of these individual local systems—will not be the same as those

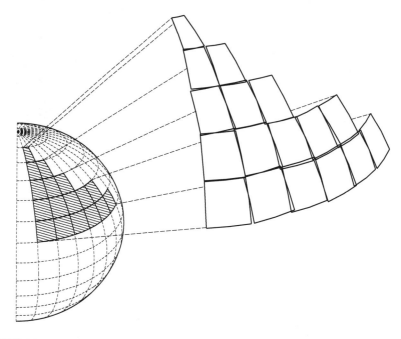

FIGURE 4.2
Local coordinate systems do not edge-match.

elements of its neighbor's coordinate system. Subsequently, there are both gaps and overlaps between adjacent maps and their attendant coordinate systems. Without a common reference system, the difficulties of moving from map to map are compounded.

So the idea of a self-consistent local map projection based on small flat planes tangent to the Earth, or the reference ellipsoid, is convenient, but only for small projects that have no need to be related to adjoining work. And as long as there is no need to venture outside the bounds of a particular local system, this method can be entirely adequate. But, generally speaking, if a significant area is involved in the work, then another strategy is needed. That is not to say that tangent-plane map projections have no larger use. For example, consider the tangent-plane map projections that are used to map the polar areas of the Earth.

Polar Map Projections

These maps are generated on a large tangent plane touching the globe at a single point, the pole. Parallels of latitude are shown as concentric circles. Meridians of longitude are straight lines from the pole to the edge of the map. The scale is correct at the center, but just as in the smaller local systems mentioned earlier, the farther you get from the center of the map, the more distorted it becomes. These maps and this whole category of map projections are called *azimuthal.* The polar aspect of two of them will be briefly mentioned: the stereographic and the gnomonic. One clear difference in their application is the position of the imaginary *light source.*

A point light source is a useful device in imagining the projection of features from the Earth onto a developable surface. The rays from this light source can be imagined to move through a translucent ellipsoid and thereby project the image of the area to be mapped onto the mapping surface, like the projection of the image from film onto a screen. This is, of course, another model for thinking about map projection, an illustration.

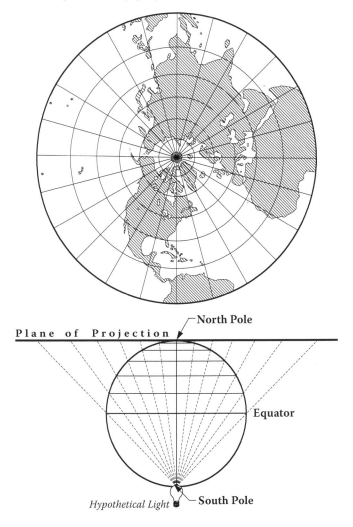

FIGURE 4.3
A stereographic projection, polar aspect.

In the case of the stereographic map projection, this point light source is exactly opposite the point of tangency of the mapping surface. In Figure 4.3, the North Pole is the point of tangency. The light source is at the South Pole. On this projection, shapes are correctly shown. In other words, a rectangular shape on the ellipsoid can be expected to appear as a rectangular shape on the map with its right angles preserved. Map projections that have this property are said to be *conformal*.

In another azimuthal projection, the *gnomonic*, the point light source moves from opposite the tangent point to the center of the globe. The term *gnomonic* is derived from the similarity between the arrangement of meridians on its polar projection and the hour marks on a sundial. The gnomon of a sundial is the structure that marks the hours by casting its shadow on those marks.

In Figure 4.3 and Figure 4.4, the point at the center of the map, the tangent point, is sometimes known as the *standard* point. In map projection, places where the map and the ellipsoid touch are known as standard lines or points. These are the only places on the map where the scale is exact. Therefore, *standard points* and *standard lines* are the only places on a map, and the resulting coordinates systems derived from them are really completely free of distortion.

As mentioned earlier, a map projection's purpose informs its design. For example, the small individual plane projections first mentioned conveniently serve work of limited scope. Such a small-scale projection is easy to construct and can support Cartesian coordinates tailored to a single independent project with minimal calculations.

Plane polar map projections are known as azimuthal projections because the direction of any line drawn from the central tangent point on the map to any other point correctly represents the actual direction of that line. And the gnomonic projection can provide the additional benefit that the shortest distance between any two points on the ellipsoid, a *great circle*, can be represented on a gnomonic map as a straight line. It is also true that all straight lines drawn from one point to another on a gnomonic map represent the shortest distances between those points. These are significant advantages to navigation on air, land, and sea. The polar aspect of a tangent plane projection is also used to augment the *Universal Transverse Mercator* projection. So there are applications for which tangent-plane projections are particularly well suited, but the distortion at their edges makes them unsuitable for many other purposes.

Decreasing that distortion is a constant and elusive goal in map projection. It can be done in several ways. Most involve reducing the distance between the map-projection surface and the ellipsoidal surface. One way this is done is to move the mapping surface from tangency with the ellipsoid and make it actually cut through it. This strategy produces a *secant* projection. A secant projection is one way to shrink the distance between the map-projection surface and the ellipsoid. Thereby the area where distortion is in an acceptable range on the map can be effectively increased, as shown in Figure 4.5.

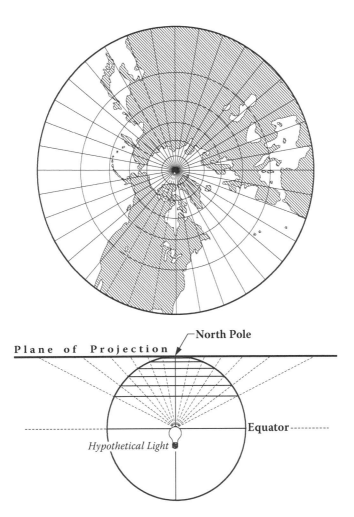

FIGURE 4.4
A gnomonic projection, polar aspect.

Another strategy can be added to this idea of a secant map-projection plane. To reduce the distortion even more, one can use one of those developable surfaces mentioned earlier, a cone or a cylinder. Both cones and cylinders have an advantage over a flat map-projection plane. They are curved in one direction and can be designed to follow the curvature of the area to be mapped in that direction. Also, if a large portion of the ellipsoid is to be mapped, several cones or several cylinders may be used together in the same system to further limit distortion. In that case, each cone or cylinder defines a *zone* in a larger coverage. This is the approach used in state plane coordinate systems.

As previously mentioned, when a conic or a cylindrical map-projection surface is made secant, it intersects the ellipsoid, and the map is brought

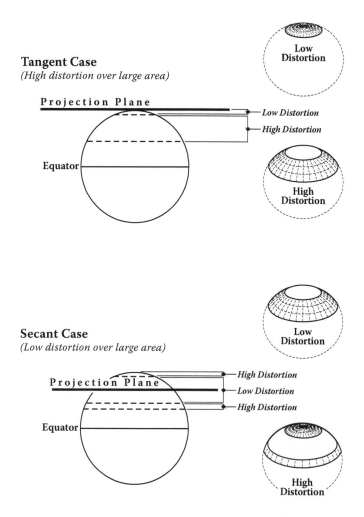

FIGURE 4.5
Minimizing distortion with a secant projection.

close to its surface. For example, the conic and cylindrical projections shown in Figure 4.6 cut through the ellipsoid. The map is projected both inward and outward onto it, and two *lines of exact scale*, standard lines, are created along the *small circles* where the cone and the cylinder intersect the ellipsoid. They are called small circles because they do not describe a plane that goes through the center of the Earth, as do the previously mentioned great circles.

Where the ellipsoid and the map-projection surface touch, in this case intersect, there is no distortion. However, between the standard lines, the map is under the ellipsoid, and outside of them, the map is above it. That means that, between the standard lines, a distance from one point to another is actually longer on the ellipsoid than it is shown on the map, and outside

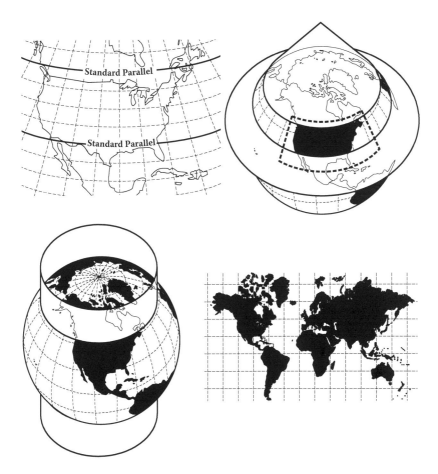

FIGURE 4.6
Secant conic and cylindrical projections.

the standard lines, a distance on the ellipsoid is shorter than it is on the map. Any length that is measured along a standard line is the same on the ellipsoid and on the map, which is why another name for standard parallels is *lines of exact scale*.

Choices

Here, and in most mapping literature, the cone and cylinder, the hypothetical light source, and other abstractions are mentioned because they are convenient models for thinking about the steps involved in building a map projection. Ultimately, the goal is very straightforward, relating each position on one surface, the reference ellipsoid, to a corresponding position on another surface as faithfully as possible and then flattening that second surface to accommodate Cartesian coordinates. In fact, the whole procedure is in the service of moving from geographic to Cartesian coordinates and back again. These days, the complexities of the mathematics are handled with computers. Of course, that was not always the case.

In 1932, two engineers in North Carolina's highway department, O. B. Bester and George F. Syme, appealed to the then Coast and Geodetic Survey (C&GS, now NGS, National Geodetic Survey) for help. They had found that the stretching and compression inevitable in the representation of the curved Earth on a plane was so severe over long route surveys that they could not check into the C&GS geodetic control stations across a state within reasonable limits. The engineers suggested that a plane coordinate grid system be developed that was mathematically related to the reference ellipsoid, but could be utilized using plane trigonometry.

Dr. Oscar Adams of the Division of Geodesy, assisted by Charles Claire, designed the first *state plane coordinate system* to mediate the problem. It was based on a map projection called the *Lambert conformal conic projection*. Dr. Adams realized that it was possible to use this map projection and allow one of the four elements of area, shape, scale, or direction to remain virtually unchanged from its actual value on the Earth, but not all four. On a perfect map projection, all distances, directions, and areas could be conserved. They would be the same on the ellipsoid and on the map. Unfortunately, it is not possible to satisfy all of these specifications simultaneously, at least not completely. There are inevitable choices. It must be decided which characteristic will be shown the most correctly, but it will be done at the expense of the others. And there is no universal best decision. Still, a solution that gives the most satisfactory results for a particular mapping problem is always available.

Dr. Adams chose the Lambert conformal conic projection for the North Carolina system. On the Lambert conformal conic projection, parallels of latitude are arcs of concentric circles, and meridians of longitude are equally spaced straight radial lines, and the meridians and parallels intersect at right angles. The axis of the cone is imagined to be a prolongation of the polar axis. The parallels are not equally spaced because the scale varies as you move north and south along a meridian of longitude. Dr. Adams decided to use this map projection in which shape is preserved based on a developable cone.

Map projections in which shape is preserved are known as *conformal* or *orthomorphic*. Orthomorphic means "right shape." In a conformal projection, the angles between intersecting lines and curves retain their original form on the map. In other words, between short lines, meaning lines under about 10 miles, a 45° angle on the ellipsoid is a 45° angle on the map. It also means that the scale is the same in all directions from a point; in fact, it is this characteristic that preserves the angles. These aspects were certainly a boon for the North Carolina highway engineers and provide benefits that all state plane coordinate users have enjoyed since. On long lines, angles on the ellipsoid are not exactly the same on the map projection. Nevertheless, the change is small and systematic. It can be calculated.

Actually, all three of the projections that were used in the designs of the original state plane coordinate systems were conformal. Each system was originally based on the North American Datum 1927 (NAD27). Along with the oblique Mercator projection, which was used on the panhandle of Alaska, the two primary projections were the Lambert conic conformal projection and the *Transverse Mercator* projection. For North Carolina, and other states that are longest east-west, the Lambert conic projection works best. State plane coordinate systems in states that are longest north-south were built on the Transverse Mercator projection. There are exceptions to this general rule. For example, California uses the Lambert conic projection even though the state could be covered with fewer Transverse Mercator zones. The Lambert conic projection is a bit simpler to use, which may account for the choice.

The Transverse Mercator projection is based on a cylindrical mapping surface much like that illustrated in Figure 4.6. However, the axis of the cylinder is rotated so that it is perpendicular to the polar axis of the ellipsoid, as shown in Figure 4.7. Unlike the Lambert conic projection, the Transverse Mercator represents meridians of longitude as curves rather than straight lines on the developed grid. The Transverse Mercator projection is not the same thing as the Universal Transverse Mercator system (UTM). UTM was originally a military system that covers the entire Earth and differs significantly from the Transverse Mercator system used in state plane coordinates.

The architect of both the Transverse Mercator projection, built on work by Gerardus Mercator, and the conformal conic projection that bears his name was Johann Heinrich Lambert, an eighteenth-century Alsatian mathematician. His works in geometry, optics, perspective, and comets are less known than his investigation of the irrationality of π. Surveyors, mappers, and cartographers know Lambert's mapping projections above all. It is especially remarkable that the projections he originated are used in every state of the United States, and both were first presented in his contribution, *Beiträge zum Gebrauche der Mathematik und deren Anwendung*, in 1772. Still, the Lambert conic projection was little used until 1918, when the U.S. Coast and Geodetic Survey, encouraged by Dr. Adams and Charles Deetz, began publishing his theory and tables from which it could be applied.

In using these projections as the foundation of the state plane coordinate systems (SPCS), Dr. Adams wanted to have the advantage of conformality and also cover each state with as few *zones* as possible. A zone in this context is a belt across the state that has one Cartesian coordinate grid, with one origin, and is projected onto one mapping surface. One strategy that played a significant role in achieving that end was Dr. Adams's use of *secant* projections in both the Lambert conic and Transverse Mercator systems.

For example, using a single secant cone in the Lambert projection and limiting the extent of a zone, or belt, across a state to about 158 miles (ca. 254 km), Dr. Adams limited the distortion of the length of lines. Not only were angles preserved on the final product, but there were also only minimal differences (for the measurement technology of the day) between the length of

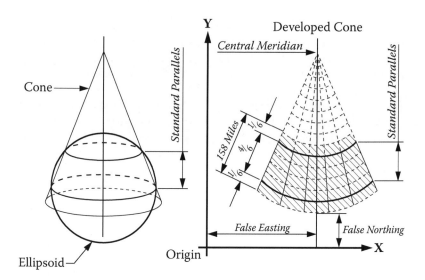

**Lambert Conformal Conic
Projection**

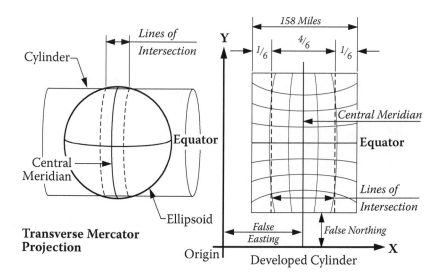

**Transverse Mercator
Projection**

FIGURE 4.7
Two SPCS map projections.

a measured line on the Earth's surface and the length of the same line on the map projection. In other words, the scale of the distortion was pretty small.

As illustrated in Figure 4.7, he placed 4/6 of the map-projection plane between the standard lines and 1/6 outside at each extremity. The distortion was held to 1 part in 10,000. A maximum distortion in the lengths of lines of 1 part in 10,000 means that the difference between the length of a 2-mile line on the ellipsoid and its representation on the map would only be about 1 ft at the most.

State plane coordinates were created to serve as the basis of a method that approximates geodetic accuracy more closely than the then commonly used methods of small-scale plane surveying. Today, surveying methods can easily achieve accuracies 1 part in 100,000 and better, but the state plane coordinate systems were designed in a time of generally lower accuracy and efficiency in surveying measurement. Today, computers easily handle the lengthy and complicated mathematics of geodesy. However, the first state plane coordinate system was created when such computation required sharp pencils and logarithmic tables. In fact, the original SPCS was so successful in North Carolina that similar systems were devised for all the states within a year or so. The system was successful because, among other things, it overcame some of the limitations of mapping on a horizontal plane while avoiding the imposition of strict geodetic methods and calculations. It managed to keep the distortion of the scale ratio under 1 part in 10,000 and preserved conformality. It did not disturb the familiar system of ordered pairs of Cartesian coordinates, and it covered each state with as few zones as possible, the boundaries of which were constructed to follow portions of county lines as much as possible, with some exceptions. The idea was that those relying on state plane coordinates could work in one zone throughout a jurisdiction.

SPCS27 to SPCS83

In Figure 4.8, the current boundaries of the SPCS zones are shown. In several instances, they differ from the original zone boundaries. The boundaries shown in the figure are for SPCS83, the state plane coordinate system based on NAD83 and its reference ellipsoid GRS80. The foundation of the original state plane coordinate system, SPCS27, was NAD27 and its reference ellipsoid, Clarke 1866. As mentioned in earlier chapters, NAD27 geographical coordinates, latitudes, and longitudes differ significantly from those in NAD83. In fact, conversion from geographic coordinates (latitude and longitude) to grid coordinates (x and y) and back is one of the three fundamental conversions in the state plane coordinate system. It is important because the whole objective of the SPCS is to allow the user to work in plane coordinates, but still have the option of expressing any of the points under consideration in either latitude and longitude or state plane coordinates without significant loss of accuracy. Therefore, when geodetic control

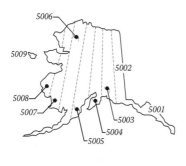

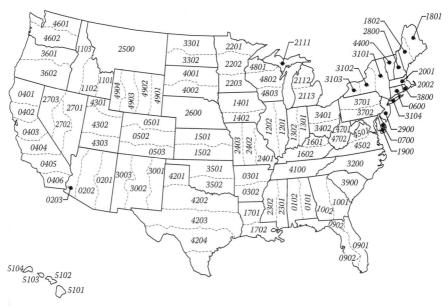

FIGURE 4.8
SPCS83 zones.

was migrated from NAD27 to NAD83, the state plane coordinate system had to go along.

When the migration was undertaken in the 1970s, it presented an opportunity for an overhaul of the system. Many options were considered, but in the end only a few changes were made. One of the reasons for the conservative approach was the fact that 37 states had passed legislation supporting the use of state plane coordinates. Nevertheless, some zones got new numbers and some of the zones changed. The zones in Figure 4.8 are numbered in the SPCS83 system known as *FIPS*. FIPS stands for Federal Information Processing Standard, and each SPCS83 zone has been given a FIPS number. These days, the zones are often known as *FIPS zones*. SPCS27 zones did not have these FIPS numbers. As mentioned earlier, the original goal was to keep each zone small enough to ensure that the scale distortion was 1 part in

10,000 or less. However, when the SPCS83 was designed, that scale was not maintained in some states.

In five states, some SPCS27 zones were eliminated altogether, and the areas they had covered were consolidated into one zone or added to adjoining zones. In three of those states—South Carolina, Montana, and Nebraska—the result was one single large zone. In SPCS27, South Carolina and Nebraska had two zones; in SPCS83, they have only one zone: FIPS zone 3900 and FIPS zone 2600, respectively. Montana previously had three zones. It now has one: FIPS zone 2500. Therefore, because the area covered by these single zones has become so large, they are not limited by the 1 part in 10,000 standard. California eliminated zone 7 and added that area to FIPS zone 0405, formerly zone 5. Two zones previously covered Puerto Rico and the Virgin Islands. They now have one: FIPS zone 5200. In Michigan, three Transverse Mercator zones were entirely eliminated.

In both the Transverse Mercator and the Lambert projection, the positions of the axes are similar in all SPCS zones. As you can see in Figure 4.7, each zone has a central meridian. These central meridians are true meridians of longitude near the geometric center of the zone. Note that the central meridian is not the y-axis. If it were the y-axis, negative coordinates would result. To avoid them, the actual y-axis is moved far to the west of the zone itself. In the old SPCS27 arrangement, the y-axis was 2,000,000 ft west from the central meridian in the Lambert conic projection and 500,000 ft in the Transverse Mercator projection. In the SPCS83 design, those constants have been changed. The most common values are 600,000 m for the Lambert conic and 200,000 m for the Transverse Mercator. However, there is a good deal of variation in these numbers from state to state and zone to zone. In all cases, however, the y-axis is still far to the west of the zone, and there are no negative state plane coordinates because the x-axis, also known as the baseline, is far to the south of the zone. Where the x-axis and y-axis intersect is the origin of the zone, and that is always south and west of the zone itself. This configuration of the axes ensures that all state plane coordinates occur in the first quadrant and are, therefore, always positive.

There is sometimes even further detail in the name of particular state plane coordinates. As refinements are made to NAD83, the new adjustments are added as a suffix to the SPCS83 label. For example, SPCS83/99 would refer to state plane coordinates that were based on a revision to NAD83 from 1999.

It is important to note that the fundamental unit for SPCS27 is the U.S. Survey foot and for SPCS83 it is the meter. The conversion from meters to U.S. Survey feet is correctly accomplished by multiplying the measurement in meters by the fraction 3937/1200.

In the following sections, the most typical conversions used in state plane coordinates will be addressed:

1. Conversion from geodetic lengths to grid lengths
2. Conversion from geographic coordinates (latitude and longitude) to grid coordinates
3. Conversion from geodetic azimuths to grid azimuths
4. Conversion from SPCS to ground coordinates

Geodetic Lengths to Grid Lengths

This brings us to the scale factor, also known as the *K factor* and the *projection factor* that the original design of the State Plane Coordinate System (SPCS) sought to limit to 1 part in 10,000. As implied by that effort, scale factors are ratios that can be used as multipliers to convert ellipsoidal lengths, also known as *geodetic distances*, to lengths on the map-projection surface; also known as *grid distances* and vice versa. In other words, the geodetic length of a line, on the ellipsoid, multiplied by the appropriate scale factor will yield the grid length of that line on the map. And the grid length multiplied by the inverse of that same scale factor would yield the geodetic length again.

While referring to Figure 4.7, it is interesting to note that on the projection used most frequently on states that are longest from east to west, i.e., the Lambert conic, the scale factor for east-west lines is constant. In other words, the scale factor is the same all along the line. One way to think about this is to recall that the distance between the ellipsoid and the map-projection surface never changes east to west in that projection. On the other hand, along a north-south line, the scale factor is constantly changing on the Lambert conic. Thus it is no surprise to see that the distance between the ellipsoid and the map-projection surface is always changing along a north-south line in that projection. But looking at the Transverse Mercator projection, the projection used most frequently on states that are longest north to south, the situation is exactly reversed. In that case, the scale factor is the same all along a north-south line, and changes constantly along an east-west line.

Both the Transverse Mercator and the Lambert conic used a secant projection surface and originally restricted the width to 158 miles. These were two strategies used to limit scale factors when the state plane coordinate systems (SPCS) were designed. Where that was not optimum, the width was sometimes made smaller, which means the distortion was lessened. As the belt of the ellipsoid projected onto the map narrows, the distortion gets smaller. For example, Connecticut is less than 80 miles wide north to south. It has only one zone. Along its northern and southern boundaries, outside of the standard parallels, the scale factor is 1 part in 40,000, a fourfold improvement over 1 part in 10,000. And in the middle of the state, the scale factor is 1 part in 79,000, nearly an eightfold improvement. On the other hand, the scale factor was allowed to get a little bit smaller than 1 part in 10,000 in Texas. By doing that, the state was covered completely with five zones. Among the guiding principles in 1933 was covering the states with as few zones as possible and having zone boundaries follow county lines. The system required ten zones and all three types of projections to cover Alaska.

In Figure 4.9, a typical 158-mile state plane coordinate zone is represented by a grid plane of projection cutting through the ellipsoid of reference. As mentioned earlier, between the intersections of the standard lines, the grid is under the ellipsoid. There, a distance from one point to another is longer on the ellipsoid than on the grid. This means that right in the middle of an SPCS zone, the scale factor is at its minimum. In the middle, a typical minimum SPCS scale factor is not less than 0.9999, though there are exceptions. Outside of the intersections, the grid is above the ellipsoid, where a distance from one point to another is shorter on the ellipsoid than it is on the grid. There, at the edge of the zone, a maximum typical SPCS scale factor is generally not more than 1.0001.

When SPCS27 was current, scale factors were interpolated from tables published for each state. In the tables for states in which the Lambert conic projection was used, scale factors change north-south with the changes in latitude. In the tables for states in which the Transverse Mercator projection was used, scale factors change east-west with the changes in x-coordinate. Today, scale factors are not interpolated from tables for SPCS83. For both the Transverse Mercator and the Lambert conic projections, they are calculated directly from equations.

There are several software applications that can be used to automatically calculate scale factors for particular stations. Perhaps the most convenient is

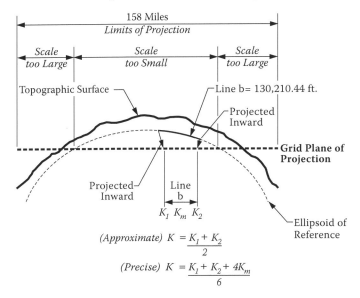

FIGURE 4.9
Scale factor.

that available free online at http://www.ngs.noaa.gov/PC_PROD/pc_prod. shtml. This is a link to the U.S. National Geodetic Survey (NGS) page where one can locate several programs that provide geodetic computational help. The program SPCS83 is available for download there. It can be used to convert latitudes and longitudes to state plane coordinates. Given the latitude and longitude of the stations under consideration, part of the available output from the program includes the scale factors for those stations. Scale factors for control stations are also available from NGS geodetic control datasheets.

To illustrate the use of these factors, consider line b in Figure 4.9 to have a length on the ellipsoid of 130,210.44 ft, a bit over 24 miles. That would be its geodetic distance. Suppose that a scale factor for that line was 0.9999536; then the grid distance along line b would be:

$$\text{geodetic distance} \times \text{scale factor} = \text{grid distance}$$
$$130{,}210.44 \text{ ft} \times 0.999953617 = 130{,}204.40 \text{ ft}$$

The difference between the longer geodetic distance and the shorter grid distance here is a little more than 6 ft. That is actually better than 1 part in 20,000; recall that the 1 part in 10,000 ratio was considered the maximum. Distortion lessens and the scale factor approaches 1 as a line nears a standard parallel.

Recall that on the Lambert projection, an east-west line, which is a line that follows a parallel of latitude, has the same scale factor at both ends and throughout. However, a line that bears in any other direction will have a different scale factor at each end. A north-south line will have a great difference in the scale factor at its north end compared with the scale factor of its south end. In this vein, note the approximate formula near the bottom of Figure 4.9:

$$K = \frac{K_1 + K_2}{2}$$

where K is the scale factor for a line, K_1 is the scale factor at one end of the line, and K_2 is the scale factor at the other end of the line. Scale factors vary with the latitude in the Lambert projection. For example, suppose the point at the north end of the 24-mile line is called Stormy and has a geographic coordinate of

37°46'00.7225" N latitude
103°46'35.3195" W longitude

and at the south end, the point is known as Seven with a geographic coordinate of

37°30′43.5867″ N latitude
104°05′26.5420″ W longitude

The scale factor for point Seven is 0.99996113, and the scale factor for point Stormy is 0.99994609. It happens that point Seven is farther south and closer to the standard parallel than point Stormy, and it therefore follows that the scale factor at Seven is closer to 1. It would be exactly 1 if it were on the standard parallel, which is why the standard parallels are called *lines of exact scale*. The typical scale factor for the line is the average of the scale factors at the two end points:

$$K = \frac{K_1 + K_2}{2}$$

$$0.99995361 = \frac{0.99996113 + 0.99994609}{2}$$

Deriving the scale factor at each end and averaging them is the usual method for calculating the scale factor of a line. The average of the two is sometimes called K_m. In Figure 4.9, there is another formula for calculating a more precise scale factor by using a weighted average. In this method, K_1 is given a weight of 1, K_m a weight of 4, and K_2 a weight of 1. No matter which method is used, the result is still called the scale factor. But that is not the whole story when it comes to reducing distance to the state plane coordinate grid. Measurement of lines must always be done on the topographic surface of the Earth, and not on the ellipsoid. Therefore, the first step in deriving a grid distance must be moving a measured line from the Earth to the ellipsoid, i.e., converting a distance measured on the topographic surface to a geodetic distance on the reference ellipsoid.

This is done with another ratio that is also used as a multiplier. Originally, this factor had a rather unfortunate name. It used to be known as the *sea level factor* in SPCS27. It was given that name because, as you may recall from Chapters 2 and 3, when NAD27 was established using the Clarke 1866 reference ellipsoid, the distance between the ellipsoid and the geoid was declared to be zero at Meades Ranch in Kansas. That meant that, in the middle of the country, the "sea level" surface, the geoid, and the ellipsoid were coincident by definition. And since the Clarke 1866 ellipsoid fit the United States quite well, the separation between the two surfaces—the ellipsoid and geoid—only grew to about 12 m anywhere in the country. With such a small distance between them, many practitioners at the time took the point of view that, for all practical purposes, the ellipsoid and the geoid were in the same place, and that place was called "sea level." Hence reducing a distance measure on the surface of the Earth to the ellipsoid was said to be reducing it to "sea level."

Today, that idea and that name for the factor is misleading because, of course, the GRS80 ellipsoid on which NAD83 is based is certainly not the same as mean sea level. Now the separation between the geoid and ellipsoid can grow as large as –53 m, and the technology by which lines are measured has improved dramatically. Therefore, in SPCS83, the factor for reducing a measured distance to the ellipsoid is known as the *ellipsoid factor*. In any case, both the old and the new names can be covered by the term *elevation factor*.

Regardless of the name applied to the factor, it is a ratio of the relationship between an approximation of the Earth's radius and that same approximation

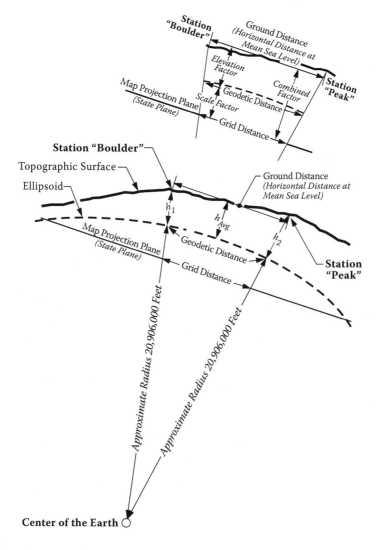

FIGURE 4.10
SPCS combined (grid) factor.

with the mean ellipsoidal height of the measured line added to it. For example, consider station Boulder and station Peak illustrated in Figure 4.10.

Boulder
   39° 59′ 29.1299″ N latitude
   105° 15′ 39.6758″ W longitude

Peak
   40° 01′ 19.1582″ N latitude
   105° 30′ 55.1283″ W longitude

The distance between these two stations is 72,126.21 ft. This distance is sometimes called the *ground distance*, or the *horizontal distance at mean elevation*. In other words, it is not the slope distance but, rather, the distance between the stations corrected to an averaged horizontal plane, as is common practice. For practical purposes, then, this is the distance between the two stations on the topographic surface of the Earth. On the way to finding the grid distance from Boulder to Peak, an interim step involves calculating the geodetic distance between them, i.e., the distance on the ellipsoid. We need the elevation factor, and here is how it is determined.

The ellipsoidal height of Boulder, h_1, is 5,437 ft. The ellipsoidal height of Peak, h_2, is 9,099 ft. The approximate radius of the Earth, traditionally used in this work, is 20,906,000 ft. The elevation factor is calculated as follows:

$$Elevation\ Factor = \frac{R}{R + h_{avg}}$$

$$Elevation\ Factor = \frac{20,906,000ft.}{20,906,000ft. + 7268ft.}$$

$$Elevation\ Factor = \frac{20,906,000ft.}{20,913,268ft.}$$

$$Elevation\ Factor = 0.99965247$$

This factor, then, is the ratio used to move the ground distance down to the ellipsoid, i.e., down to the geodetic distance.

Ground distance Boulder to Peak = 72,126.21 ft
Geodetic distance = ground distance × elevation factor
Geodetic distance = 72,126.21 × 0.99965247
Geodetic distance = 72,101.14 ft

It is possible to refine the calculation of the elevation factor by using an average of the actual radial distances from the center of the ellipsoid to the end points of the line, rather than the approximate 20,906,000 ft. In the area of stations Boulder and Peak, the average ellipsoidal radius is actually a bit longer, but within the continental United States, such variation will not cause a calculated geodetic distance to differ significantly. However, it is worthwhile to take care to use the ellipsoidal heights of the stations when calculating the elevation factor, rather than the orthometric heights.

In calculating the elevation factor in SPCS27, no real distinction is made between ellipsoidal height and orthometric height. However, in SPCS83, the averages of the ellipsoidal heights at each end of the line can be used for h_{avg}. If an ellipsoidal height is not directly available, it can be calculated from the formula

$$h = H + N$$

where
   h = ellipsoidal height
   H = orthometric height
   N = geoidal height

As mentioned previously, converting a geodetic distance to a grid distance is done with an averaged scale factor:

$$K = \frac{K_1 + K_2}{2}$$

In this instance, the scale factor at Boulder is 0.99996703, and at Peak it is 0.99996477.

$$0.99996590 = \frac{0.99996703 + 0.99996477}{2}$$

Using the scale factor, it is possible to reduce the geodetic distance 72,101.14 ft to a grid distance:

geodetic distance × scale factor = grid distance
72,101.14 ft × 0.99996590 = 72,098.68 ft

There are two steps: first from ground distance to geodetic distance using the elevation factor and second from geodetic distance to grid distance using the scale factor. These two steps can be combined into one. Multiplying the elevation factor and the scale factor produces a single ratio that is usually known as the *combined factor* or the *grid factor*. Using this grid factor, the measured line is converted from a ground distance to a grid distance in one jump.

Here is how it works. In the example considered here, the elevation factor for the line from Boulder to Peak is 0.99965247 and the scale factor is 0.99996590:

$$\text{grid factor} = \text{scale factor} \times \text{elevation factor}$$
$$0.99961838 = 0.99996590 \times 0.99965247$$

Then, using the grid factor, the ground distance is converted to a grid distance

$$\text{grid distance} = \text{grid factor} \times \text{ground distance}$$
$$72{,}098.68 \text{ ft} = 0.99961838 \times 72{,}126.21 \text{ ft}$$

The grid factor can also be used to go the other way. If the grid distance is divided by the grid factor, it is converted to a ground distance.

$$\text{Ground Distance} = \frac{\text{Grid Distance}}{\text{Grid Factor}}$$

$$72{,}126.21 \text{ ft} = \frac{72{,}098.68 \text{ ft}}{0.99961838}$$

Geographic Coordinates to Grid Coordinates

Consider again two previously mentioned stations, Stormy and Seven. Stormy has an NAD83 geographic coordinate of

$$37° \ 46' \ 00.7225'' \text{ N latitude}$$
$$103° \ 46' \ 35.3195'' \text{ W longitude}$$

and Seven has an NAD83 geographic coordinate of

$$37° \ 30' \ 43.5867'' \text{ N latitude}$$
$$104° \ 05' \ 26.5420'' \text{ W longitude}$$

Finding the state plane coordinates of these stations can be accomplished online. The NGS Web site (http://www.ngs.noaa.gov/TOOLS) provides several programs, and among them is one named *state plane coordinates*. It has two convenient routines that allow the user to convert a station from SPCS to latitude and longitude, and vice versa, in both NAD27 and NAD83 and to discover its scale factor and convergence angle. It is interesting to note that the site also includes free routines for conversion from SPCS27 to SPCS83. The conversion is a three-step process. The SPCS27 coordinate is converted to an NAD27 geographic coordinate, a latitude and longitude. Next the NAD27 geographic coordinate is transformed into an NAD83 geographic coordinate, and finally the NAD83 geographic coordinate becomes an SPCS83 coordinate. This procedure is common to nearly all GIS (geographic information system) software. In any case, using the state plane coordinates software, it is possible to find the SPCS83 coordinates for these two stations:

Stormy: N 428,305.869

             E 1,066,244.212

Seven: N 399,570.490

             E 1,038,989.570

Both of these are in meters, the native units of SPCS83. The original SPCS27 design was based on the use of the U.S. Survey foot as its unit of measurement. That remains the appropriate unit for that system today. However, SPCS83 is a bit more complicated in that regard. While the fundamental unit for SPCS83 is the meter, when it comes to converting coordinates from meters to feet, one of two conversions is called for. One is the conversion from meters to U.S. survey feet. The other conversion is from meters to international feet. The international foot—so named because an international agreement was established to define 1 in. equal to 2.54 cm exactly—is equal to 0.3048 m. That version of the foot was adopted across the United States, except at the U.S. Coast and Geodetic Survey (C&GS). In 1959, instead of forcing C&GS to refigure and republish all the control station coordinates across the country, it was given a reprieve under a notice in the *Federal Register* (24 FR 5348). The agency was allowed to retain the old definition of the foot, which is the U.S. Survey foot, in which 1 m is 39.37 in., exactly. Another way of saying it is a U.S. Survey foot is 1200/3937 m.

It was decided that C&GS could continue to use the U.S. Survey foot until the national control network was readjusted, but following the adjustment, the agency was to switch over to the international foot. That did not quite work out. When NAD83 was fully established, the agency that had replaced C&GS (the NGS) mandated that the official unit of all the published coordinate values would be the meter. Then in 1986, NGS declared it would augment its publication of state plane coordinates in meters with coordinates for the same stations in feet. The question is: which foot? The answer is: the version legislated by the state in which the station was found.

In practical terms, this means that in states such as the 11 that chose the U.S. Survey foot and the 6 that chose the international foot, it is clear which should be used. However, 14 states do not specify the version of the foot that is official for their SPCS. The remaining 19 states have no legislation on the state plane coordinates at all.

The difference between the two units for the foot may seem an academic distinction, but consider the conversion of station Stormy. It happens that the station is in the Colorado 0503 or South Zone, and that state has chosen U.S.

Survey feet. It follows that the correct coordinate values for the station in U.S. Survey feet are:

Stormy: N 1,405,200.17
           E 3,498,169.55

However, if the metric coordinates for Stormy were mistakenly converted to international feet, they would be N 1,405,202.98 and E 3,498,176.55. The distance between the correct coordinate and the incorrect coordinate is more than 7½ ft, with the largest difference occurring in the easting.

Conversion from Geodetic Azimuths to Grid Azimuths

As was mentioned in Chapter 1, meridians of longitude converge, and that fact has an effect on the directions of lines in SPCS. To illustrate the rate of that convergence, consider the east and west sides of a 1 square mile figure somewhere in the middle of the conterminous United States. Suppose that the two sides were both meridians of longitude on the surface of the Earth and the directions of both lines were the same, geodetic north. However, if that square-mile figure were projected onto a Lambert conic or Transverse Mercator SPCS, the directions would no longer be equal. Their azimuths would suddenly differ by about 1 min of arc. And unless one of the sides happens to follow the central meridian of the zone, neither of their azimuths would be grid north.

In SPCS, the direction known as grid north is always parallel to the central meridian for the zone. The east and west lines of the square mile in the example follow meridians on the surface of the Earth. Meridians converge at the pole; they are not parallel to one another. In SPCS, north is grid north, and the lines of the grid are parallel to each other. They must also be parallel to one another and the central meridian of the zone, so geodetic north and grid north are clearly not the same. In fact, only on that central meridian do grid north and geodetic north coincide. Everywhere else in the zone they diverge from one another, and there is an angular distance between them. In SPCS27, that angular distance was symbolized with the Greek letter theta (θ) in the Lambert conic projection and by delta alpha ($\Delta\alpha$) in the Transverse Mercator projection. However, in SPCS83, convergence is symbolized by gamma (γ) in both the Lambert conic and the Transverse Mercator projections. In both map projections, east of the central meridian grid north is east of geodetic north, and the convergence angle is positive. West of the central meridian grid, north is west of geodetic north, and the convergence angle is negative.

The angle between geodetic and grid north, the convergence angle, grows as the point gets further from the central meridian. It also gets larger as the latitude of the point increases. The formula for calculating the convergence in the Transverse Mercator projection is

$$\gamma = (\lambda_{cm} - \lambda) \sin \varphi$$

where
λ_{cm} = the longitude of the central meridian
λ = the longitude through the point
φ = the latitude of the point

The formula for calculating the convergence angle in the Lambert conic projection is very similar; it is

$$\gamma = (\lambda_{cm} - \lambda) \sin \varphi_0$$

where
λ_{cm} = the longitude of the central meridian
λ = the longitude through the point
φ_0 = the latitude of the center of the zone

As an example, here is the calculation of the convergence angle for station Stormy. It is in the Colorado South Zone on a Lambert conic projection, where the longitude of the central meridian is longitude 105° 30′ 00″ W and φ_0 is latitude 37° 50′ 02.34″ N

$\gamma = (\lambda_{cm} - \lambda) \sin \varphi_0$
$\gamma = (105° \ 30′ \ 00″ - 103° \ 46′ \ 35.3195″) \sin 37° \ 50′ \ 02.34″$
$\gamma = (01° \ 43′ \ 24.68″) \sin 37° \ 50′ \ 02.34″$
$\gamma = (1° \ 43′ \ 24.68″) \ 0.6133756$
$\gamma = +1° \ 03′ \ 25.8″$

The angle is positive, as expected, east of the central meridian.
The formula used to convert a geodetic azimuth to a grid azimuth includes the convergence angle and another element:

grid azimuth = geodetic azimuth − convergence angle + second term

Another way of stating the same formula is

$$t = \alpha - \gamma + \delta$$

in which
t = grid azimuth
α = geodetic azimuth
γ = the convergence angle
δ = the second term

The second term (see Figure 4.11) is included because lines between stations on the ellipsoid are curved, and that curvature is not completely eliminated when the geodetic azimuth line is projected onto the SPCS grid.

Several factors affect the extent of the curve of the projected geodetic azimuth line on the grid. The direction of the line and the particular map projection from which the grid was created are two of those elements. For instance, a north-south line does not curve at all when projected onto the Lambert conic grid, and neither does an east-west line when projected onto the Transverse Mercator grid. However, in both cases, the more a line departs from these cardinal courses, the more it will curve on the grid. In fact, the maximum curvature in each projection occurs on lines that are parallel to their standard lines. That means that an east-west line would have the largest curve in a Lambert conic projection, and a north-south line would have the largest in a Transverse Mercator projection.

Another factor that affects the size of the curve is the distance of the line from the center of the zone. In a Transverse Mercator projection, the farther a line is from the central meridian, the more it will curve. In the Lambert conic projection, the farther a line is from the central parallel of latitude through the center of the zone, φ_0, the more it will curve. Finally, in both map projections, the longer the line, the more it will curve. It follows, therefore, that long

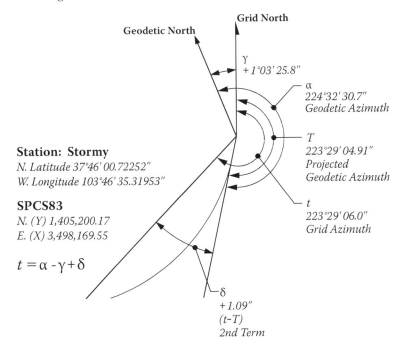

Grid North

Geodetic North

γ
$+1°03'25.8''$

α
$224°32'30.7''$
Geodetic Azimuth

T
$223°29'04.91''$
Projected
Geodetic Azimuth

t
$223°29'06.0''$
Grid Azimuth

Station: Stormy
N. Latitude 37°46'00.72252''
W. Longitude 103°46'35.31953''

SPCS83
N. (Y) 1,405,200.17
E. (X) 3,498,169.55

$t = \alpha - \gamma + \delta$

δ
$+1.09''$
$(t-T)$
2nd Term

FIGURE 4.11
Second term.

lines at the boundaries between zones depart the most from straight lines on the grid. A 2-mile north-south line in a Transverse Mercator SPCS will deviate about 1 arc-sec from a straight line. In the Lambert conic SPCS, an east-west line of that length will deviate about 1 arc-sec from a straight line.

Even though the distance between station Seven and station Stormy is approximately 24 miles, the departure from a chord between the two is only about 1″ at Stormy and 2″ at Seven (see Figure 4.12). Note, however, that the grid bearing of the line between the two is approximately N 43½° E rather than east-west.

This correction for curvature is sometimes known as *t-T*, the *arc-to-chord* correction, and the *second term*. The correction is small and pertinent to the most precise work. However, it is important to note that, like the convergence itself, the second term comes into play only when there is a need to convert a conventionally observed azimuth into a grid azimuth in SPCS. Where optical surveying data is used, that will almost certainly be required. On the other hand, if the work is done with GPS observations, or where field observations are not involved at all, convergence and the second term are not likely to affect the calculation of the work.

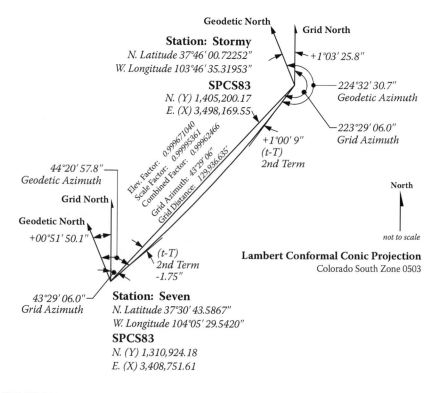

FIGURE 4.12
Stations Stormy and Seven.

SPCS to Ground Coordinates

When a state plane coordinate is assigned to a station on the Earth, the coordinate is not really on the station itself. As described previously, the coordinate is on a grid. The grid is most often below—but sometimes above—the actual location of the point it intends to represent. Simply put, the point is on the Earth, but the coordinate is not. This rather inconvenient fact can be, and sometimes is, ignored. However, it is valuable to remember that while SPCS is generally designed to provide scale distortion no worse than 1 part in 10,000 or so, modern measurement is routinely better than that. It is therefore common for users of the system to find that their measurement from one place to another turns out to be longer or shorter than an inverse between the SPCS coordinates indicates. Such results sometimes lead to an assumption that the whole system is shot through with unacceptable inaccuracies. That is certainly not the case. It is also usual that, once the actual cause of the difference between ground and grid dimensions is fully understood, a convenient resolution is sought. Frequently, the resolution is bringing the state plane coordinates from the grid to the ground.

This is done by extending the idea mentioned at the end of the section on conversion from geodetic lengths to grid lengths. The idea is to divide the grid distance by the grid (or combined) factor to find the ground distance. This concept is sometimes applied to more than the single line where the particular grid factor is actually correct. It is used to convert many lines and many points over an entire project. In those instances, a single grid factor calculated near the center of a project is used. This one grid factor is intended to convert grid distances between points to the ground distances, ignoring the changes in the scale factor and the elevation factor from point to point.

The coordinates that result from this approach are not state plane coordinates, and they are often truncated to avoid being confused with actual SPCS. The typical truncation is dropping the first two digits from the northings and eastings of these project coordinates. This strategy in some ways defeats the purpose of the SPCS, which—when used correctly—offers a reasonably accurate approximation of geodetic positions that is consistent over a large area. However, fixing one grid factor for a project returns a user to the sort of tangent-plane system mentioned earlier. Such a project coordinate system cannot be joined along its edges with a neighboring system without difficulty. This difficulty cannot be avoided because the plane created near the elevation of the center of the project inevitably departs more and more from the reference ellipsoid at the edges. The advantages of a large secant map-projection plane are removed. Therefore, as long as one stays within the now-local system, the work can progress smoothly, but outside of the area it will not match other systems.

It is interesting to note that some governmental organizations have established localized projections, often at the county level, to bring grid coordinates closer to the ground. The objective has been to provide a coordinate basis that is more convenient for building a local GIS.

Common Problems with State Plane Coordinates

As mentioned earlier, the official native unit of SPCS83 coordinates is the meter. However, reporting in feet is often required. Many states prefer U.S. Survey feet: Nebraska, Wyoming, Colorado, California, Connecticut, Indiana, Maryland, North Carolina, and Texas. Other states specify international feet: Arizona, Michigan, Montana, Oregon, South Carolina, and Utah. Still others have taken no official action on the issue. Nevertheless, clients in any state may request coordinates in either format.

If an error is suspected in converting SPCS83 coordinates from meters to feet, look to the easting of the coordinate. Since the false easting in SPCS is quite large, it is there that the discrepancy will be most obvious, as illustrated in the example conversion of station Stormy.

Another common problem stems from the periodic readjustments performed by NGS. As mentioned in Chapter 2, NAD83 has been subject to refinements since it replaced NAD27. These improvements are largely due to the increasing amount of GPS information available and are denoted with a suffix, such as NAD83/91, the latter number referring to the year of the readjustment. Since SPCS83 is based on NAD83, these adjustments result in new state plane coordinates as well. It is therefore feasible that one county may use say NAD83/86 coordinates and an adjoining county may use NAD83/94 coordinates. The result may be a different coordinate assigned to the same station, both in NAD83, but differing as the result different adjustments. The solution is to take care with the year of adjustment when using published coordinates.

When there is a discrepancy of millions of meters or feet between the eastings of coordinates of points that are certainly not hundreds of miles apart, the error may be attributable to SPCS27 coordinates among SPCS83 coordinates, or vice versa. There were substantial changes made to the false easting when the datums were changed. Another possible culprit for the condition could be the combination of coordinates of one SPCS83 zone with coordinates from the adjoining zone.

UTM Coordinates

A plane coordinate system that is convenient for GIS work over large areas is the Universal Transverse Mercator (UTM) system. UTM combined with the Universal Polar Stereographic (UPS) system covers the world in one consistent system. In terms of scale, it is four times less accurate than typical state-plane coordinate systems. Yet the ease of using UTM and its worldwide coverage make it very attractive for work that would otherwise have to cross many different SPCS zones. For example, nearly all National Geospatial-Intelligence Agency (NGA)—formerly National Imagery and Mapping Agency (NIMA) and formerly the Defense Mapping Agency—topographic maps, USGS quad sheets, and many aeronautical charts show the UTM grid lines.

It is often said that UTM is a military system created by the U.S. Army. In fact, several nations and the North Atlantic Treaty Organization (NATO) played roles in its creation after World War II. At that time, the goal was to design a consistent coordinate system that could promote cooperation among the military organizations of several nations. Before the introduction of UTM, the Allies found that their differing systems hindered the synchronization of military operations.

Conferences were held on the subject from 1945 to 1951 with representatives from Belgium, Portugal, France, and Britain, and the outlines of the present UTM system were developed. By 1951, the U.S. Army had introduced a system that was very similar to that currently used.

The UTM projection divides the world into 60 zones. Actually, one could say that there are 120 zones, since each of the 60 zones is divided into a Northern Hemisphere portion and Southern Hemisphere portion at the equator. The numbering of the zones begins at longitude 180° E, the International Date Line, and increases sequentially toward the east. Zone 1 is from 180° W longitude to 174° W longitude, zone 2 is from 174° W longitude to 168° W longitude, and so on. The conterminous United States are within UTM zones 10 to 19 (see Figure 4.13). Each zone has a central meridian exactly in the middle. For example, in Zone 1 from 180° W longitude to 174° W longitude, the central meridian is 177° W longitude, so each zone extends 3° east and west from its central meridian. The central meridian for zone 2 is 171° W longitude.

Here is a convenient way to find the zone number for a particular longitude. Consider west longitude negative and east longitude positive, add 180°, and divide by 6. Any answer greater than an integer is rounded to the next highest integer, and you have the zone. For example, Denver, Colorado, is near 105° W Longitude, −105°.

$$-105° + 180° = 75°$$

$$75°/6 = 12.50$$

Round up to 13

Therefore, Denver is in UTM zone 13.

All UTM zones have a width of 6° of longitude (see Figure 4.14). From north to south, the zones extend from from 84° N latitude to 80° S latitude. Originally, the northern limit was at 80° N latitude, and the southern limit was at 80° S latitude. On the south, the latitude is a small circle that conveniently traverses the ocean well south of Africa, Australia, and South

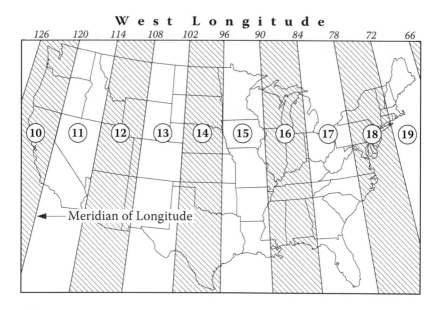

$\boxed{10}$ = UTM zone number

FIGURE 4.13
UTM zones in the conterminous United States.

America. However, 80° N latitude was found to exclude parts of Russia and Greenland and was extended to 84° N latitude.

These zones nearly cover the Earth, except the polar regions, which are covered by two azimuthal zones called the Universal Polar Stereographic (UPS) projection. The foundation of the UTM zones is a secant Transverse Mercator projection very similar to those used in some state plane coordinate systems. The UTM secant projection covers approximately 180 km between the lines of exact scale, where the cylinder intersects the ellipsoid. The scale factor grows from 0.9996 along the central meridian of a UTM zone to 1.00000 at 180 km to the east and west. Recall that SPCS zones are usually limited to about 158 miles and, therefore, have a smaller range of scale factors than the UTM zones. In state plane coordinates, the scale factor is usually no more than 1 part in 10,000. In UTM coordinates, it can be as large as 1 part in 2,500.

The reference ellipsoids for UTM coordinates vary among five different figures, but in the United States the reference is the Clarke 1866 ellipsoid. However, one can obtain 1983 UTM coordinates by referencing the UTM zone constants to the GRS80 ellipsoid of NAD83.

As mentioned earlier, unlike any of the systems previously discussed, every coordinate in a UTM zone occurs twice, once in the Northern Hemisphere and once in the Southern Hemisphere. This is a consequence

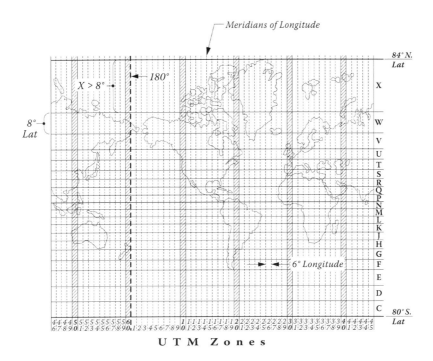

FIGURE 4.14
UTM zones around the world.

of the two origins in each UTM zone. The origin for the portion of the zone north of the equator is moved 500 km west of the intersection of the zone's central meridian and the equator. This arrangement ensures that all of the coordinates for that zone in the Northern Hemisphere will be positive. The origin for the coordinates in the Southern Hemisphere for the same zone is also 500 km west of the central meridian, but it is not at its intersection with the equator; it is 10,000 km south of it, close to the South Pole. This orientation of the origin guarantees that all of the coordinates in the Southern Hemisphere are in the first quadrant and are positive. In both hemispheres and for all zones, the easting (the x-value) of the central meridian is 500,000 m at the central meridian, as shown in Figure 4.15. The developed UTM grid is defined in meters.

In fact, in the official version of the UTM system, there are actually more divisions in each UTM zone than the north-south demarcation at the equator. As shown in Figure 4.14, each zone is divided into 20 subzones. Each of the subzones covers 8° of latitude and is lettered from C on the south to X on the north. Actually, subzone X is a bit longer than 8°. Remember the extension of the system from 80° N latitude to 84° N latitude: That all went into subzone X. It is also interesting that I and O are not included. They resemble 1 and 0 too closely.

There are free utilities online to convert UTM coordinates to latitude and longitude at http://www.ngs.noaa.gov/cgi-bin/utm_getgp.prl and another to convert latitude and longitude to UTM coordinates at http://www.ngs.noaa.gov/cgi-bin/utm_getut.prl, both courtesy of NGS.

A word or two about the polar zones that round out the UTM system. The Universal Polar Stereographic (UPS) projections are conformal azimuthal stereographic projections like those mentioned earlier. The projections

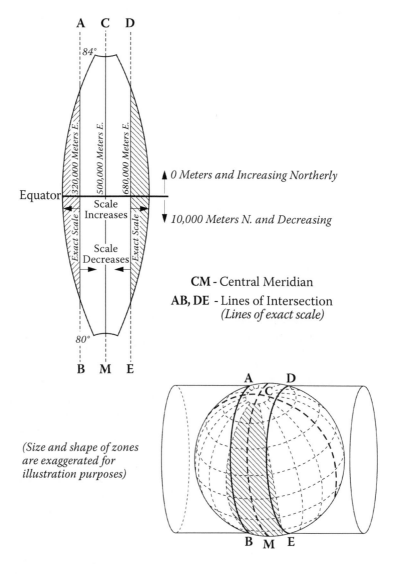

FIGURE 4.15
A UTM zone.

have two zones. The north zone covers the North Pole and the south zone the South Pole. The projections are on a plane tangent at a pole and perpendicular to the minor axis of the reference ellipsoid. The projection lines originate from the opposite pole. As shown in Figure 4.16, the minimum scale factor is 0.994 at the pole. It increases to 1.0016076 at 80° latitude from each pole. The scale factor is constant along any parallel of latitude. The line of exact scale is at 81° 06′ 52.3″ N latitude at the North Pole and 81° 06′ 52.3″ S latitude at the South Pole. In both cases, the pole is given a false easting and northing: The *x*-coordinate (easting) of the pole is 2,000,000 m,

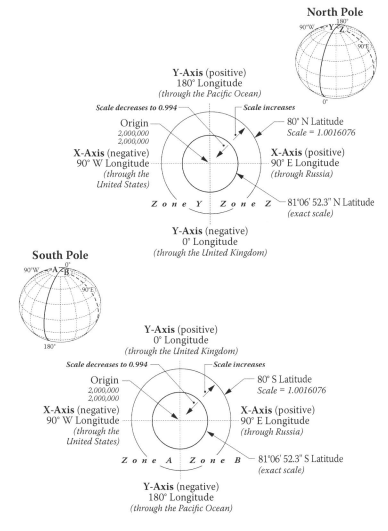

FIGURE 4.16
The Universal Polar Stereographic (UPS) projection.

and the y-coordinate (northing) of the pole is 2,000,000 m. The reference ellipsoid at both the North and South Poles is the International Ellipsoid.

At the North Pole and at the South Pole, the x-axis lies along the meridians 90° W and 90° E . The easting values start at 2,000,000 m at the pole and increase along the 90° E meridian. At the North Pole, the y-axis lies along the meridians 0° and 180°, and the northings start from 2,000,000 m at the pole and increase along the 180° meridian. At the South Pole, the y-axis also lies along the meridians 0° and 180°, and the northings start from 2,000,000 m at the pole but increase along the 0° meridian.

Exercises

1. Which of the following general statements about the state plane coordinate systems in the United States is incorrect?

   a. There are three map projections used in SPCS in the United States.

   b. The map projections used in the SPCS cause meridians of longitude to appear as straight lines on the SPCS grid.

   c. Some states utilize more than one map projection in their SPCS.

   d. The axis of the Lambert cone is coincident with the Earth's polar axis; the axis of the Transverse Mercator cylinder is perpendicular to it.

2. Which of the following statements does not properly describe a change in the state plane coordinate systems from SPCS27 to SPCS83?

   a. There have been changes in the distances between the SPCS zones and their origins.

   b. SPCS83 coordinates are expressed in more units than were SPCS27 coordinates.

   c. There are fewer SPCS83 zones than there were SPCS27 zones in some states.

   d. In SPCS83, the scale factor limit of 1 part in 10,000 has been exceeded for the first time.

3. Which of the following statements correctly describes an aspect of conformal map projection in SPCS?

   a. It ensures that the angle between two lines on the map-projection surface is the same as it is on the reference ellipsoid, regardless of the length of the lines.

   b. In a conformal map projection, the scale is the same in all directions from a point.

   c. Convergence in a conformal map projection is always much less than 1°.

   d. The Lambert conic projection is a conformal projection, but the Transverse Mercator projection is not.

4. Which of the following statements about the scale factor in SPCS is true?

   a. The scale factor changes constantly along the central meridian of a Transverse Mercator SPCS zone.

   b. Scale factor ratios are 1 part in 10,000 in SPCS zones.

   c. A scale factor is used in SPCS to convert a length on the reference ellipsoid to a length on the map-projection plane grid.

   d. The scale factor does not change along the central meridian of a Lambert conic SPCS zone.

5. If a line on a secant map-projection plane crossed one of the standard lines, what would be the difference between the scale factor at the end between the standard lines and the scale factor at the end outside the standard lines?

   a. The scale factor at the station outside the standard lines would be greater than 1, and the scale factor for the station between the standard lines would be less than 1.

   b. For an east-west line that crossed a standard parallel in a Lambert conic projection, the scale would be the same at both ends of the line.

   c. For a north-south line that crossed a standard line of exact scale in a Transverse Mercator projection, the scale would be the same at both ends of the line.

   d. The scale factor at the station outside the standard lines would be less than 1, and the scale factor for the station between the standard lines would be greater than 1.

6. Which of the following statements about convergence angles in SPCS is not correct?

   a. As the distance from the central meridian increases, the absolute values of the convergence angles increase.

   b. As the length of a line connecting two stations increases, the absolute values of the convergence angles at each end station increase.

   c. As the direction of a line connecting two stations approaches the cardinal course, north-south or east-west, the absolute values of the convergence angles at each end station are not affected.

   d. As the latitude decreases, the absolute values of the convergence angles decrease.

7. Which of the following statements about the second-term, also known as t-T and arc-to-chord, correction in SPCS is not correct?

   a. As the distance from the central meridian in a Transverse Mercator projection and the distance from the central latitude, φ_0, in the Lambert conic projection increases, the absolute values of the second-term angles increase.

   b. As the length of a line connecting two stations increases, the absolute values of the second-term angles at each end station increase.

   c. As the direction of a line connecting two stations becomes more nearly parallel with a zone's standard lines, the absolute values of the second-term angles at each end station increase.

   d. As the latitude decreases, the absolute values of the second-term angles decrease.

8. The combined factor is used in SPCS conversion. How is the combined factor calculated?

   a. The scale factor is multiplied by the elevation factor.

   b. The scale factor is divided by the grid factor.

   c. The grid factor is added to the elevation factor and the sum is divided by 2.

   d. The scale factor is multiplied by the grid factor.

9. Which value is closest to the typical change in the false easting between SPCS27 and SPCS83?

   a. The typical difference between the false easting in the Lambert conic is about 6 miles, and the typical difference in the Transverse Mercator is about 30 miles.

   b. The typical difference between the false easting in the Lambert conic is about 1,400,000 m, and the typical difference in the Transverse Mercator is about 3,000,000 m.

   c. The typical difference between the false easting in the Lambert conic is about 9600 ft, and the typical difference in the Transverse Mercator is about 4800 ft.

   d. The typical difference between the false easting in the Lambert conic is about 47,600 ft, and the typical difference in the Transverse Mercator is about 31,500 ft.

10. Which of the following does not correctly state a difference between the Transverse Mercator coordinate system used in SPCS and the Universal Transverse Mercator (UTM) coordinate system?

   a. The UTM system of coordinates was designed and established by the U.S. Army in cooperation with NATO member nations. The Transverse Mercator SPCS was established by the U.S. Coast and Geodetic Survey.

   b. The official unit of measurement in the UTM system of coordinates is the meter. The official unit in Transverse Mercator SPCS83 is the international foot.

   c. The UTM system of coordinates is augmented by the Universal Polar Stereographic (UPS) system to complete its worldwide coverage. The Transverse Mercator SPCS does not have such a broad scope.

   d. The scale factor in the UTM system of coordinates can reach 1 part in 2,500. In the Transverse Mercator SPCS, the scale factor is usually no more than 1 part in 10,000.

Explanations and Answers

1. Explanation:

Every state, Puerto Rico, and the U.S. Virgin Islands has its own plane rectangular coordinate system. The state plane coordinate systems are based on the Transverse Mercator projection, an oblique Mercator projection, and the Lambert conic map-projection grid.

Generally speaking, states with large east-west extent use the Lambert conic projection. This system uses a projection cone that intersects the ellipsoid at standard parallels. The axis of the cone is imagined to be a prolongation of the polar axis of the Earth. When the cone is developed, that is, opened to make a plane, the ellipsoidal meridians become straight lines that converge at the cone's apex. The apex is also the center of the circular lines that represent the projections of the parallels of latitude.

Some states use both the Lambert conic and the Transverse Mercator projections for individual zones within the state system. Some rely on the Transverse Mercator projection alone. The Transverse Mercator projection uses a projection cylinder whose axis is imagined to be parallel to the Earth's equator and perpendicular to its axis of rotation. It intersects the ellipsoid along standard lines parallel to a central meridian. However, after the cylinder is developed, all the projected meridians and parallels become curved lines.

Answer: **(b)**

2. Explanation:

In both SPCS27 and SPCS83, eastings are reckoned from an axis placed far west of the coordinate zone so that they all remain positive. Northings are reckoned from a line far to the south for the same reason, from a baseline south of the zone. Placing the y-axis 2,000,000 ft west from the central meridian in the Lambert conic projection and 500,000 ft west in the Transverse Mercator projection was usual in SPCS27. Those distances are now often 600,000 m for the Lambert conic and 200,000 m for the Transverse Mercator in SPCS83. There is some variation in these SPCS83 distances, but there are no negative state plane coordinates in either SPCS27 or SPCS83 because the design of both ensures that all coordinates fall in the first quadrant.

The native unit for SPCS27 coordinates is U.S. Survey feet. The native unit for SPCS83 coordinates is meters. SPCS83 coordinates are also expressed in either U.S. Survey feet or international feet, depending on the state.

In five states, some SPCS27 zones were eliminated altogether, and the areas they had covered were consolidated into one zone or added to adjoining zones. In three of those states, the result was one single large zone. Those states are South Carolina, Montana, and Nebraska. The large single zones caused the 1 part in 10,000 scale factor standard to be exceeded in these states. However, it was not the first time. The standard had previously been exceeded by SPCS27 in several zones.

Answer: **(d)**

3. Explanation:

Both the Lambert conic and the Transverse Mercator projections used in SPCS are conformal. *Conformality* means that an angle between short lines on the ellipsoid is preserved after it is mapped onto the map-projection plane. This feature allows the shapes of small features to look the same on the map as they do on the Earth. Map projections can be equal area, azimuthal, equidistant, or conformal. It would be great if one map projection could simultaneously possess all of these desirable characteristics at once, but that, unfortunately, is not possible. Conformality was built into the design of the SPCS from the beginning. It allows the transformation of points in such a way that, for all practical purposes, the angles between short lines on the projection surface are the same as they are on the reference ellipsoid because the scale is the same in all directions from a point. However, as the lengths of the lines grow, a difference in the angles between them does become significant. This difference is an artifact of the curvature of lines on the ellipsoid, and a correction known variously as the t-T, the arc-to-chord correction, or the second term comes into play. Conformality does not limit the convergence of meridians, which is a function of the size of the SPCS zone.

Answer: **(b)**

4. Explanation:

The ratio of the length between two stations on the map-projection plane and that same length on the reference ellipsoid is the scale factor. When the gap between the map-projection plane and the reference ellipsoid does not change along the length of a line, the scale factor does not change. When the gap between the two surfaces varies along the length of a line, the scale factor varies. Therefore, on the conic Lambert projection, the scale factor is constant along east-west lines and changes continuously along north-south lines. Since the central meridian in a conic projection is north-south, and the gap between the map-projection plane and the reference ellipsoid varies along such a line, the scale factor varies. On the cylindrical Transverse Mercator projection, the scale factor is constant along north-south lines and changes continuously along east-west lines. The central meridian in a cylindrical projection is north-south, and the gap between the map-projection plane and the reference ellipsoid does not vary along such a line; the scale factor is constant.

It is not correct to say that the scale factor for SPCS zones has strictly adhered to the design criterion of 1 part in 10,000. For example, of those states that utilized the Lambert conic projection in SPCS27, there were 8 of the more than 70 zones across the United States in which the scale factor ratio was smaller than 1 part in 10,000 along their central meridians. The situation has changed somewhat in SPCS83; for example, Montana, Nebraska, and South Carolina now have one zone each, instead of several, with the attendant change in the scale factor. Although the objective limit was originally 1 part in 10,000, in fact, it has not been maintained universally as the SPCS system has evolved.

Answer: **(c)**

5. Explanation:

In a perfect map projection, the scale factor would be 1 everywhere. Since that is not possible, in a secant projection the scale factor is less than 1 between standard lines and greater than 1 outside the standard lines. The only place where the scale factor is 1 in such a projection is along the standard lines themselves. The larger the variation in scale factors that is allowed in the original design of the system, the larger their range above and below 1.

Neither an east-west line in the Lambert conic projection nor a north-south line in the Transverse Mercator projection would cross a standard line.

Answer: **(a)**

6. Explanation:

In both the Transverse Mercator and Lambert conic projections, the absolute value of the convergence angle increases with the distance from the central meridian. It also increases with the latitude. In Lambert conic zones, the effect of latitude is constant for the zone, but it is greater in more northerly zones. In Transverse Mercator zones, the effect of the latitude grows constantly from south to north. Since convergence angles are tied to the position of the stations on the grid, they are not affected by the length or the direction of the lines connecting those stations.

Answer: (b)

7. Explanation:

In both projections, Transverse Mercator and Lambert conic, the absolute values of the second-term corrections at each end of a line are at their maximum when the line is at the edge of a zone and parallel to the standard lines. A north-south line does not exhibit any curve at all when projected onto the Lambert conic grid and neither does an east-west line when projected onto the Transverse Mercator grid. However, in both cases, the more a line departs from these cardinal courses, the more it will curve on the grid. That means that an east-west line would have the largest curve in a Lambert conic projection, and a north-south line would have the largest in a Transverse Mercator projection. Again, in a Transverse Mercator projection, the farther a line is from the central meridian, the more it will curve. In the Lambert conic projection, the farther a line is from the central parallel of latitude through the center of the zone, φ_0, the more it will curve. A longer line will have a larger second-term correction at its ends than will a shorter line.

The second-term correction is usually small and ignored for much of the work. Its characteristics are essentially the same in SPCS27 and SPCS83. While there is a relationship between the position of a line on a particular SPCS zone and the second-term correction, there is none between the latitude of the line and the size of the second-term correction.

Answer: (d)

8. Explanation:

The grid factor changes with the ellipsoidal height of the line. It also changes with its location in relation to the standard lines of its SPCS zone. The grid factor is derived by multiplying the scale factor by the elevation factor. The product is nearly 1 and is known as either the grid factor or the combined factor. There is a different combined factor for every line in the correct application of SPCS.

Answer: (a)

9. Explanation:

To avoid negative coordinates in SPCS, the *y*-axis was moved west of the zone itself. In the SPCS27, the typical false easting from the *y*-axis to the central meridian was 2,000,000 ft (378.7879 miles) in the Lambert conic projection and 500,000 ft (94.6970 miles) in the Transverse Mercator projection. There were substantial changes made to the false easting when the datums were changed. The most common values in SPCS83 are 600,000 m (327.822 miles) for the Lambert conic projection, a shift of about 6 miles. The typical SPCS83 Transverse Mercator false easting is 200,000 m (124.274 miles), a shift of about 30 miles. However, there is a good deal of variation in these numbers from state to state and zone to zone.

Answer: **(a)**

10. Explanation:

Both the UTM system and the Transverse Mercator system in SPCS use the meter as their official unit of measurement.

Answer: **(b)**

5

Rectangular System

Some assert that the *Public Land Survey System (PLSS)* is a coordinate system. Others strongly maintain that it is certainly not. Suffice it to say that for more than 200 years it has been, and continues to be, a system to divide land and describe property across most of the United States. It was originally established by the Land Ordinance Act of May 20, 1785. The system has been incrementally improved through practical innovation and government regulation since then. The Department of the Treasury had jurisdiction over the Public Land Survey System from its beginnings until 1812. It was a sensible arrangement, since the government was poor in cash and rich in land. The land came to be known as the *public domain,* meaning land that was once, or is even now, owned by the federal government of the United States. What was required was a practical, reliable, and unambiguous method of disposal. The fledgling federal government needed to do several things. It needed to pay soldiers, often with land instead of cash. It needed to disperse the population on the land to realize its potential. Put simply, it needed to raise money.

From the very beginning of the system, it has always been the policy of the federal government that land in the public domain must be surveyed before it is sold to others, i.e., before it is *patented*. Therefore, the United States embarked on the most extensive surveying project ever undertaken. The work began in Ohio in the autumn of 1785. On April 25, 1812, the Congress created the *General Land Office (GLO)* to survey and administer the PLSS. Under the GLO, the work was done through a contract system. The contractors were deputy surveyors who worked for surveyors general across the country. The surveying was not actually done by government employees until 1910. Later, the Bureau of Land Management was created in the Department of the Interior in 1946. It has been in charge of the system since then.

The PLSS was and is an eminently practical attempt to cover a large portion of the curved surface of the Earth with a Cartesian grid with 1-mile squares, *sections*. The degree to which it has succeeded can be most clearly seen when flying over the central and western United States. The orderly patchwork quality of the landscape is a testament to those who created and those who maintain the rectangular system.

Leaving aside the question of whether it is a coordinate system or not, the PLSS belongs in this book because of its prominent place in the definition of the lands and the geographic information systems (GIS) in the 30 states created from the public domain. These lands were acquired by purchase, treaty, or cession, and they once amounted to over 1.8 billion acres. The majority of that land has now been mapped into sections and passed from the federal government into other hands, but its measurement is still, in most cases, governed by the rules of the PLSS.

The Public Land Surveying System is a huge subject. The small portion of it that will be presented here is intended to offer a basic understanding of a few elements of the topic that are pertinent to practitioners of GIS. Even though the application of the system can be complex, the design has always been straightforward and practical. It is that design, as currently practiced, that will be emphasized here.

From the beginning, the PLSS has been adaptable. It had to be. It needed to apply to a variety of conditions on the land and be capable of changing with the times. It also needed to be as unambiguous as possible to prevent misunderstandings and boundary disputes. It has achieved these goals remarkably well.

Initial Points

In a sense, *initial points* are the origins of the PLSS. They represent the intersection of two axes, as seen in Figure 5.1. There are 32 initial points in the conterminous United States and 5 in Alaska. Initial points were first mentioned in the *Manuals of Instructions* in 1881. There were several of these manuals. The first official version was issued by the GLO in 1851. They were the vehicles by which the commissioner of the GLO communicated the methods of survey to the surveyors general. They reflect the evolution of the system's surveying procedures. However, by the time the establishment of initial points was mentioned in a manual, many of the 37 in place today had already been set. Nevertheless, the 1881 manual stated:

> Initial points from which the lines of the public surveys are to be extended must be established whenever necessary under such special instructions as may be prescribed by the Commissioner of the General Land Office. The locus of such initial points must be selected with great care and due consideration for their prominence and easy identification, and must be established astronomically. (GLO 1881, 35)

In other words, the points were set somewhat arbitrarily, as needed, and were assigned latitudes and longitudes derived astronomically.

East and west from the initial point, a *baseline* was laid out. This baseline was intended to follow an astronomically determined parallel of latitude. And it did so as nearly as possible, with corner monuments established every half mile or every *40 chains*. The chain was, and is, the native unit of the PLSS

Detail - Ohio

FIGURE 5.1
Initial points of the federal system of rectangular surveys.

system. A Gunter's chain is 66 ft long, and therefore, 40 chains equal 2640 ft or half a mile. It is also important to note that 10 square chains equal 1 acre, a convenient relationship.

A *principal meridian* was extended north and south from the initial point following a meridian of longitude, again astronomically determined, on which corner monuments were set every half mile, every 40 chains. While the initial point and the baseline are not usually named, it is common for the principal meridian to have a name, such as the Wind River Principal Meridian, the Ute Principal Meridian, or the 6th Principal Meridian.

These two lines—the baseline and the principal meridian—are the fundamental axes and the foundation from which the PLSS was extended across the country. They are by no means abstract historical curiosities. By their monumentation, these lines serve as a real physical presence today. And the baselines and principal meridians are geographical lines, each as nearly a parallel of latitude and a meridian of longitude as the practical measurement technology of their day could achieve.

However, it should be borne in mind that much of the early surveying in the PLSS was done with a solar compass and a linked chain. The discrepancies between the design and the reality of the monumented corners may be larger than would be expected if the work were done today with modern equipment. Despite that caveat, the original work was remarkably good. Whether good or bad, however, there is no question it takes precedence over subsequent retracements or resurveys. The monuments set during the execution of the original survey, and the boundary lines they describe, are correct and inviolable by law. It is an important principle of the Public Land Survey System and is mentioned, among other places, in a section of a federal statute originally enacted in 1925, 43 Stat. 1144, which says:

> All the corners marked in the surveys returned by the Secretary of the Interior or such agency as he may designate, shall be established as the proper corners of the sections or subdivision of sections they were intended to designate.... The boundary lines, actually run and marked in the surveys returned by the Secretary of the Interior or such agency as he may designate, shall be established as the proper boundary lines of the sections, or subdivisions, for which they were intended. (BLM 1973, 6)

It is possible to think that, since most of the public domain is no longer under the jurisdiction of the federal government, the quoted statute does not necessarily apply to lands in private hands. However, the laws and courts in most states that were created from public domain follow the line that the rules established by federal statutes and elaborated in the various manuals of instructions will hold sway over the lands in state and private ownership under their jurisdiction.

At the intersection of the principal meridian and the baseline, there is an initial point that is also represented by an actual monument. In fact, the *Manual of Instructions of 1902* not only stipulated that initial points should be set in conspicuous locations that are visible from a distance, but that they should be perpetuated by indestructible monuments like a copper bolt set in a rock ledge. Monuments perpetuate the initial points; most have been remonumented, but the original monuments of 8 of the 37 initial points are still standing. The baselines and principal meridians that extend in cardinal directions from them often, though not always, terminate at a state line. Once these axes were in place, the next step was the creation of *quadrangles*.

Quadrangles

Quadrangles in the PLSS are large rectangular areas bounded by meridians of longitude and parallels of latitude. They are 24 miles on a side by current design, though that has not always been the dimension used. At each initial point, a principal meridian on one side and a baseline on another bound four such quadrangles, and on the other two sides there is a *guide meridian* and a *standard parallel,* as illustrated in Figure 5.2. The lines around a quadrangle are known as *standard lines.*

A guide meridian was intended to follow a meridian of longitude. A standard parallel was intended to follow a parallel of latitude. And they both were established astronomically with corner monuments set every 40 chains, each half mile, just as was done to create the principal meridian and the baseline. Over a broad area of the PLSS, only one baseline and one principal meridian originate from one initial point, but there are many quadrangles built from them. These standard lines are numbered as you see in Figure 5.2. At a distance of 24 miles north from the baseline is the first standard parallel north, and 24 miles south from the baseline is the first standard parallel south. Twenty-four miles east from the principal meridian is the first guide meridian east, and 24 miles west from the principal meridian is the first guide meridian west. This logical system of numbering was and is carried through

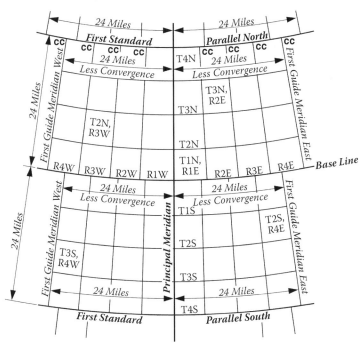

cc = Closing Corner

FIGURE 5.2
Survey of quadrangles.

the whole region governed by a particular principal meridian and baseline forming 24-mile quadrangles bounded by geographic, standard lines.

The reason for this approach was explained in the *Manual of Instructions of 1890*:

> Standard parallels shall be established at intervals every 24 miles, north and south of the base line, and guide meridians at intervals of every 24 miles east and west of the Principal Meridian; the object being to confine the errors resulting from the convergence of meridians, and inaccuracies in measurements, within the tracts of lands bounded by the lines so established. (GLO 1890, 18)

Since the guide meridians followed astronomically derived meridians of longitude, they converged. Even though the first guide meridian east may have been 24 miles east of the principal meridian at the baseline, by the time it has run north for 24 miles and intersected the first standard parallel north, that guide meridian was considerably closer to the principal meridian than when it started the trip. And convergence was unavoidable because the principal meridian and the first guide meridian, the second guide meridian, the third, and so on, were all meridians of longitude. Therefore, as described in the 1890 manual, and illustrated in Figure 5.2, the guide meridians were not then and are not today continuous. They were designed to stop at each standard parallel and the baseline so that they could be corrected back to their original 24-mile spacing.

At the south end of each of its segments, a guide meridian was intended to be 24 miles from the principal meridian, or neighboring guide meridian. But at its north end, where it terminated at the standard parallel or baseline, the spacing between meridians was inevitably less than 24 miles due to convergence of the meridians. As a consequence, that segment of the guide meridian stopped there. A new segment was begun, but before that next segment of the guide meridian proceeded north for the next 24 miles, it was shifted back to the correct spacing at the standard parallel to again be 24 miles from the principal meridian or its neighboring guide meridian. In this way, it was ensured that the south end of each segment of a guide meridian was then and is today 24 miles east or west from its neighboring guide meridian at its south end. But the north end of each segment of a guide meridian was and is less than 24 miles from the next guide meridian because of convergence. And this is the reason that standard parallels are also known as *correction lines*. They are the lines at which guide meridians are corrected for convergence.

It is sensible, then, that the process of actually laying out guide meridians proceeded from the south to north. Corner monuments were set every 40 chains along the way, as done with the principal meridian and the baseline. When the guide meridian finally intersected the standard parallel, a *closing corner* monument was set where a surveyed boundary, here the guide meridian, intersected a previously established boundary, here the standard parallel or baseline. Any excess or deficiency in the measurement that had accumulated in the work along the guide meridian was placed in the last half mile preceding its intersection with the east-west standard line. So the last half mile at the north end of a segment of a guide meridian may be more or less than 40 chains. But more importantly, the cardinal direction of the guide meridian was maintained all the way to its intersection with the standard parallel or baseline. And therefore that intersection could not possibly fall on the already established corner monument on the parallel it was intersecting, because the guide meridian was no longer at the 24-mile spacing it had when it started.

Before the guide meridian was run, a *standard corner* had been set on the standard parallel or baseline 24 miles from the principal meridian or guide meridian to serve as a corner of that quadrangle. However, that standard corner could only be the correct corner of the quadrangle to the north, because when the guide meridian was closed on the standard parallel or baseline, the intersection of the two lines certainly fell somewhat west or east of the standard corner because of convergence. As a result, there was a standard corner and a closing corner at every corner of a quadrangle, and they were some distance from each other, with the exception of quadrangle corners set on the principal meridian including the initial point. In every other case, there was a closing corner monument set. It was and is the corner for the quadrangles to the south. The corner monument for the quadrangle to the north will usually be a standard corner.

Townships

The quadrangle then became the framework enclosing the land from which 16 *townships* could be created (see Figure 5.3). Townships are the units of survey of the PLSS. A township is approximately 6 miles on a side and is designed to include 36 sections. The boundaries of a township are intended to follow meridians of longitude and latitudinal lines.

Please recall that, as the standard lines around the quadrangle were run, corner monuments were set at increments of half a mile, 40 chains. Some of those corner monuments were destined to become *township corners*. These were set every 6 miles, 480 chains, along the standard lines, which are the boundaries of the quadrangle. From these corners, the townships were established.

In the actual surveying of township, the *meridional*, north-south, boundaries had precedence. They were surveyed from south to north through the 24-mile block of land, the quadrangle. These north-south township boundary lines were laid out along meridians of longitude as nearly as possible. As the meridional lines were laid out, the corner monuments were set every 40 chains, every half mile. On the other hand, the latitudinal boundaries of the

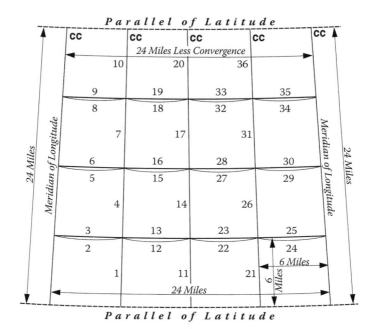

cc = Closing Corner

FIGURE 5.3
The order of running the lines in the subdivision of a quadrangle into townships.

township, the east-west lines, were first run on a trial or *random* line from east to west on most township lines. Random lines were also known as *blank* lines. Corner monuments were only set every 40 chains as the latitudinal line was corrected to a *true* line, usually west to east in these cases.

This random and true line method was used when corner monuments were in place at both ends of a boundary, but the boundary had not yet been actually connected by a surveyed line. It worked this way: A trial line, i.e., the random line, was created when the surveyors pushed their survey west from the township corner monument they had just established while surveying north up the meridian, say, at the northeast corner of a township. The next objective was to survey to the corner monument they knew had already been set 6 miles away on the western boundary of the township, its northwest corner. As the surveyors proceeded toward it on the random line, they set temporary points every 40 chains. When finally they intersected the western boundary of the township, most likely they missed their objective corner. When that happened, the distance north or south that this random line missed the northwest corner was known as the *falling*. The falling was very instructive; it told the surveyors the distance they needed to correct the random line so that it would actually connect the northwest and northeast corner monuments with a true latitudinal boundary. So, the surveyors would return east on the corrected line, this time from the northwest township corner to the northeast township corner. And along the way they would correct the temporary points they had set on the random line and set the actual corner monuments on the true line every 40 chains, except for those on the westernmost half mile.

All excess or deficiency of the 6-mile measurement was placed in the westernmost half mile. For example, suppose that the surveyors found that, instead of 480 chains from the northeast to the northwest corners of the township, they measured 479.80. The design of the PLSS dictated that the deficiency of 0.20 chains is not to be distributed evenly through the 6-mile line. It was and is all absorbed in the final half mile. The result is that 11 of the distances between the corner monuments on the true line were 40 chains, as expected, and the 12th, the westernmost, was 39.80 chains. In this way, there were always as many regular sections in a township as possible.

Here is how the procedure was described in the *Manual of Instructions 1902*:

130. Whenever practicable, the township exteriors in a block of land 24 miles square, bounded by standard lines, will be surveyed successively through the block, beginning with those of the southwestern township.
131. The meridional boundaries of the township will have precedence in the order of survey and will be run from south to north on true meridians, with permanent corners at lawful distances; the latitudinal boundaries will be run from east to west on random or trial lines, and corrected back on true lines. The falling of a random, north or south of the township corner to be closed upon, will be carefully measured.
132. When running random lines from east to west, temporary corners will be set at intervals of 40.00 chains, and proper permanent corners will be established upon the true line, corrected back in accordance with these instructions, thereby throwing the excess and deficiency against the west boundary of the township as required by law. (GLO 1902, 57)

The order of the running of the lines to create the townships out of the quadrangle is illustrated in Figure 5.3. It began at the southeast corner of the southwestern township. The first line run was the eastern boundary of that township, which was terminated with the setting of the northeast township corner. It is a meridional boundary. These north-south township lines are also known as *range lines*. The next line was a random line along the northern latitudinal boundary of the first *tier* of townships. After the random line was corrected for the falling, it was returned on a true line back to the northeast township corner. Then the range line was continued northward, the next township corner was set, and random and true lines were run to the township corner on the standard line. The procedure was repeated on the next township boundaries north, and finally the range line was closed on the northern parallel of the quadrangle. That closing corner, labeled CC in Figure 5.3, was established in much the same way as the northeast and northwest corners of the quadrangle described earlier. And once again, the excess and deficiency in the measurement of the range line were placed in the northernmost half mile, that may have contained more or less than 40 chains.

The surveying of the township boundaries continued, and the pattern was repeated for the next range line. It was also begun at its south end. Corner monuments were set every 40 chains for 6 miles as the meridional line was surveyed northward. Once again, the surveyor ran a random line west, but this time the objective corner was not on a standard line, but on the recently set northeast township corner. The falling was noted, and the line was returned on the true line, as corner monuments were set every 40 chains. These procedures were continued until the range line was closed on the northern quadrangle boundary. The last range line in the quadrangle required a variation. It was begun on its south end, like the others. After being run north for 6 miles, a random line and a true line were run west, like the others. But after the true line was returned, another random line was run from the township corner east to the eastern boundary of the quadrangle and returned on the true line. Here, too, the excess and deficiency of the 6-mile measurement were absorbed in the westernmost 40 chains.

When the subdivision of a quadrangle into four tiers and four ranges of townships was done according to the design of the PLSS, 16 townships whose boundaries were very nearly meridians of longitude and parallels of latitude were created. The inevitable errors in the east-west distance measurements were taken up in the last half mile on the west side of each township. The errors in the north-south measurements were thrown into the last half mile on the north side of the northern tier of townships, just south of the north boundary of the quadrangle. And all four boundaries of all 16 townships had corner monuments in place about every 40 chains.

In fact, such clarity was seldom achieved. There were almost always some extenuating circumstances that interrupted the orderly execution of the standard PLSS plan. Quadrangles that contained large or navigable rivers or other large bodies of water could not be surveyed so neatly. Quadrangles that contained previous bona fide rights in land outside the purview of PLSS, such as patented mineral claims or Spanish land grants, required that the orderly pattern of running lines be interrupted. In these cases, the normal lines were closed upon the boundaries of the senior property rights or the limits of the water as defined by a *meander* line. (The meander line is discussed in the Fractional Lots section of this chapter.) In fact, there were myriad difficulties that frequently prevented straightforward subdivision of a quadrangle into townships. Nevertheless, the fundamental plan was most often carried out much as described. When difficulties did arise, the surveyors involved did their utmost to use procedures, however unorthodox, that created a subdivision that fit the design.

Sections

The township then became the framework around the land from which 36 sections could be created. Since 1796, the numbering of the sections has begun in the northeast corner of the township. It has been sequential from east to west in the first tier of sections, 1 through 6. Section 7 has been south from section 6, and the second tier has been sequential from west to east, 7 through 12. Section 13 has been south from section 12, with the third tier sequential from east to west again. This zigzag continues until section 36 in the southeast corner of the township. This pattern is *boustrophedonic*, that is, it alternates from right to left and then left to right. It is said that the system harks back to an ancient mode of writing and the path an ox takes plowing a field. Whatever the origin of the scheme, the same basic plan is also used on the numbering of *fractional lots* within the sections. (Fractional lots are discussed in greater detail in the Fractional Lots section.)

Sections are the units of subdivision of the PLSS. They are approximately 1 mile on a side and contain approximately 640 acres (see Figure 5.4). Since 1890, the boundaries of a section have not been intended to follow meridians of longitude and parallels of latitude. Section lines are intended to be parallel to the *governing boundaries* of the township. The governing boundaries are the east and the south township boundaries, unless those township boundaries are defective.

Even though the procedure for the present method of subdividing a township into sections was fairly well established by 1855, here is a description of the procedure from the *Manual of Instructions of 1930*.

> The subdivisional survey will be commenced at the corner of sections 35 and 36, on the south boundary of the township, and the line between sections 35 and 36 will be run parallel to the east boundary of the township ... establishing the corner of sections 25, 26, 35 and 36. From the last named corner, a random line will be run eastward ... parallel to the south boundary of section 36 to its intersection with the east boundary of the township, placing 40 chains from the beginning a post for temporary quarter-section corner.... If the random line intersects said township boundary to the north or south of said corner, the falling will be calculated and the true line ... established.... The meridional section line will be continued on the same plan, likewise the latitudinal section lines. After having established the west and north boundaries of section 12, the line between sections 1 and 2 will be projected northward on a random line parallel to the east boundary of the township ... to its intersection with the north boundary of the township.... If ... said random line intersects the north boundary of the township to the east or west of the corner of sections 1 and 2 the falling will be carefully measured and from the data thus obtained the true course returned course will be calculated. (GLO 1930, 181)

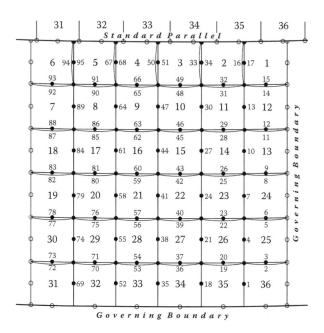

o Standard corner monuments set when township exterior was run.

⊸ Closing corner monuments set when township exterior was run.

• Standard corner monuments set during subdivision of township.

➤ Closing corner monuments set during subdivision of township.

FIGURE 5.4
The order of running the lines for the subdivision of a township into sections.

In other words, when a township was subdivided, the eastern range of sections was first in line. The work began at the southwest corner of section 36 and the line was run parallel with the eastern boundary of the township. A corner monument halfway, at 40 chains, was called a *quarter* corner, and a corner monument set at 80 chains from the beginning was the section corner. Quarter corners are so named because connecting these monuments with straight cardinal lines across a section divides it into quarters,

Next, the boundary between sections 36 and 25, a latitudinal line, was run on a random line, headed for the objective corner at the northeast corner of section 36. Once the falling was discovered, the line was corrected and run back, with the final quarter corner monument set halfway. This procedure was repeated all the way up to the meridional line between sections 1 and 2. At this point, there were two possibilities.

If the northern line of the township was not a standard parallel, the north-south line was run on a random line toward the objective corner that had been set when the north line of the township had been laid out. A temporary quarter corner was set at 40 chains from the south, with the excess and deficiency of the measurements thrown into the last half mile. Once the falling was found to the objective corner, the true line returned southward, and the temporary quarter corner was replaced with a permanent monument, still 40 chains north of the section corner to the south.

On the other hand, if the northern line of the township was a standard parallel, the north-south line was still parallel with the east township boundary. However, in this case, there was not an objective corner. Please recall that in this circumstance, the monuments that had been set when the north line of the township, the standard parallel, was laid out were only applicable to the townships north of the standard parallel, because of convergence of meridians. Nevertheless, during the subdivision of the township into sections, a quarter corner was set at 40 chains from the south, with the excess and deficiency of the measurements thrown into the last half mile. A closing corner was set at the intersection of the meridional section line with the standard parallel, and the distance between the closing corner and the previously set corners on each side of it was recorded.

Using this plan, a township could be fully subdivided. A deviation from the procedure occurred when the last meridional line was run starting between sections 31 and 32. The random and true line method was used to establish the latitudinal lines to both the east and west.

The final result of the work was a township that contained 36 sections. Working with sections today, it is clear that most are nearly a mile square. But those along the northern and western boundaries of the township are not so regular, nor were they intended to be. The design of the system itself pushes the distortion from convergence and the accumulated measuring errors into the last half mile of the tier of sections along the north and the range of sections along the west. It is there that the excesses and deficiencies are absorbed, but only after as many regular sections as possible have been carved out of the township.

Creating as many regular sections as possible has been an objective of the PLSS from the beginning, but the rectangular system has had to be flexible. The surface of the Earth does not easily accommodate such a regular geometry. Applying the plan to lands adjoining rivers and lakes—areas where there may be an incomplete south or east township boundary—has called for ingenuity. These *fractional* townships often made it impossible to follow the general rules laid down in the manuals of instructions. Joining the townships and sections from one initial-point region with those from another and the irregularity of defective township exteriors also tend to disrupt the design.

Subdivision of Sections

As mentioned earlier, straight lines—run from the quarter-section corner monuments that stand in the middle of the four section lines—divide the section into quarters. This method was stipulated in the Act of February 11, 1805. The intersection of these lines is considered the center of the section, or center quarter corner, and is a corner common to the four quarter sections. Government surveyors set the quarter-section corner monuments when they created the sections inside the township, but generally they did not subdivide the sections. Running the lines not set by the government surveyors, including the subdivision of sections, has always been left to the local surveyors. Other acts of Congress in the early nineteenth century provided definitions of the half-quarter and quarter-quarter portions of a section, the Acts of 1820 and 1832, respectively. However, actually creating those parts on the ground was again left to local surveyors.

These quarter and half divisions are known as *aliquot* parts of a section. A division of something is an aliquot when the divisor is an integer and there is no remainder. But in practical terms, this has meant that an aliquot part of a section in the PLSS is a half or a quarter of the previously larger subdivision. These are its legal subdivisions. The PLSS does not recognize other divisions such as thirds or fifths. For example, quarter sections can be further divided into halves and quarters, and they are legal subdivisions as well. The quarter-quarter section, i.e., a quarter of a quarter section, is the smallest legal subdivision per the *Manual of Instruction of 1973*. The 1973 manual is the most current version of that long series. In other words, a normal section contains approximately 640 acres. The typical aliquot parts of such a section are four. Half sections contain approximately 320 acres. Quarter sections contain approximately 160 acres. Half-quarter sections contain approximately 80 acres. Quarter-quarter sections contain approximately 40 acres.

Township Plats

When the notes of the original surveys of the townships in the public lands were returned to the GLO, and later the Bureau of Land Management (BLM), plats were developed from the work. These plats represent the township included in the survey. They show the direction and length of each line and their relation to the adjoining surveys. They also include some indication of the relief, the boundaries, and the descriptions and area of each subdivision of the sections. As mentioned previously, the government will not convey lands in the public domain to others until the survey has been done and the official plat has been filed and approved. This is true, in part, because the deed (the patent) that eventually grants the ownership of the land does so with direct reference to the township plat. It is an integral part of the transaction and binds the parties to the specific dimensions of the land as shown on the township plat. Therefore, the orientation and size of the parts of the sections on the township plat are significant. Also significant, then, is the creation of the lines on the township plat by *protraction*.

Protraction, in this context, refers to drawing lines on the township plat that were not actually run or monumented on the ground during the official subdivision of the land. For example, the lines that divide the sections into quarters—the quarter lines—are shown on the township plat as dashed lines. They are protracted because, even though the quarter-corner monuments were set, the official government surveyors never ran the quarter lines themselves. Rules were drafted to guide the local surveyors who would eventually run those lines. Rules stipulated that the lines must be straight from quarter corner to quarter corner, and their intersection must be the center of the section, but the government was not responsible for running the lines and simply drew them on the plat. The same may be said of the lines that divided the quarter sections into quarters, the quarter-quarter sections. They are also protracted lines and were not run on the ground by the GLO or BLM, except under special instructions.

Therefore, it follows that, for all practical purposes, the aliquot parts of a section—virtually all of the lines and the areas shown inside a section on an official township plat—were created by protraction. They were drawn on the plat, not run on the ground. This partition includes the boundaries of another legal subdivision of a section that is not an aliquot at all. It is known as a fractional lot.

Fractional Lots

Lotting occurs in fractional sections (see Figure 5.5), and the definition of a fractional section has grown a bit over the years. GLO Commissioner George Graham wrote one of the earliest definitions of the term to George Davis in 1826:

> It is here proper to premise that the technical meaning of "fractional section" is a tract of land not bounded by sectional lines on all sides in consequence of the intervention of a navigable stream or some other boundary recognized by law and containing a less quantity than six hundred & forty acre. (White 1983, 83)

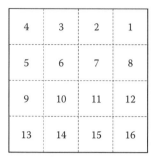

Fractional Lot Numbering Plan

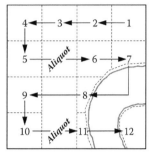

Fractional Lot Numbering in Practice

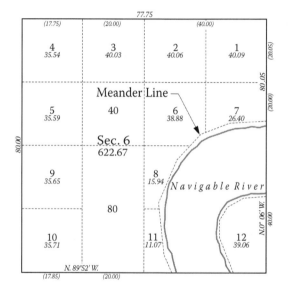

FIGURE 5.5
Fractional lots.

This definition has expanded somewhat over the years. It still applies to the sections the commissioner mentioned, but it also includes the sections on the north and west of a normal township. Please recall that the excess and deficiency of the measurements made during the subdivision of a quadrangle into townships and the subdivision of a township into sections is absorbed in the last half mile along the northern and western boundaries. By design, that means that the last half of those sections is something more or less than 40 chains (2640 ft).

One of the principles of the PLSS is always to create as many regular aliquot parts as possible. In practice, this means that half of that last half mile of a section on the northern and western boundaries of a township can be held to 20 chains. Doing that creates the last regular aliquot part before hitting the north or west boundary of the township. However, the remainder of that distance is then to become a side of a parcel that cannot be described as an aliquot part because it is not, nor can it be, a half or a quarter of the previously larger subdivision. Therefore, it is known as a fractional lot.

All along the north and the west boundary of a normal township, one will find one row or more of these fractional lots. The sections bordering the north and west boundaries of the township, except section 6, are normally subdivided into two regular quarter sections, two regular half-quarter sections, and four fractional lots. In section 6, the subdivision will show one regular quarter-section, two regular half-quarter sections, one regular quarter-quarter section, and seven fractional lots. This is the result of the plan of subdivision.

However, that is by no means the only place that fractional lots are found. Fractional lots are found anywhere a section has an irregular boundary, or as Commissioner Graham wrote, they comprise any section that does not have complete boundaries on all sides—for example, a section invaded by a body of water. In such a case, a *meander line* segregates the water, and the section or sections around it are fractional.

A meander line is a line run along the mean high-water mark of a permanent natural body of water. The meander line generally follows the curves, twists and turns of the shore. It is run to determine the area of the land in a fractional section, after the area covered by the water is excluded. Since its inception, the PLSS has been concerned with transferring land from the public domain to others, and therefore the area of usable property contained in the aliquot parts of sections has been a matter of keen interest. However, a navigable river is not a part of the public domain. It must forever remain a common public highway and is excluded from the system. Under current instructions, a body of water is meandered if it is a navigable river, a stream more than 3 chains wide, or a lake larger than 50 acres. It is important to note that the meander line around the water is not actually a property boundary; nevertheless, closing corners are set on meander lines. When a standard line, township line, section line, or aliquot line intersects a meander line, a *meander corner* is set to perpetuate the intersections. Where that happens, fractional lots spring up. They are protracted against meander lines.

The same thing happens when a more recent PLSS survey collides with a former official government survey. For example, it is inevitable that PLSS surveys originating at one initial point will grow outward and eventually meet the PLSS surveys that originate from another initial point. When that occurs, the lines and areas of the two systems will certainly not match. Fractional lots are protracted where they join. When the PLSS survey encounters boundaries around property that existed before the rectangular system was used, fractional lots are protracted again. This is the case when lines are closed on a reservation, national park, mining claim, previous land grant, agreement, or other bona fide private claim. In each of these instances, the guiding principle is that fractional sections should be subdivided in a way that produces as many aliquot parts as possible.

Fractional lots are numbered. The system of their numbering follows the same model used to number sections. It is a zigzag pattern that begins at the most northeasterly lot, which is Lot 1. Then, continuing within the same section, the next lot to the west is Lot 2. It is important to note that the next lot is not necessarily immediately adjacent to Lot 1. The general plan is that Lot 1 through Lot 4 are numbered east to west sequentially, as shown in Figure 5.5. The numbering scheme then drops down to the next tier of lots south from Lot 4 to Lot 5. Then the plan would be to sequentially number lots from west to east, Lot 5 through Lot 8, drop south to the next tier and number the lots east to west, Lot 9 through Lot 12. Finally, the southernmost tier would be numbered west to east Lot 13 through Lot 16. However, there are not necessarily fractional lots in each tier, and aliquot parts are not numbered. Therefore, in a typical section 6, the lots on the northern tier are numbered east to west 1 through 4, and Lot 5 is south of Lot 4 according to plan. But from Lot 5 east there are usually aliquot parts, no lots. Lot 6 then is numbered south of Lot 5 because it is simply the next fractional lot encountered by the zigzag numbering system. In other words, the scheme can be imagined to proceed east from Lot 5 seeking lots to number, but none are there. It then drops south to the next tier, still none. It then proceeds east to west, and finally along the western boundary encounters a fractional lot that then becomes Lot 6. The same thing happens between Lot 6 and Lot 7.

The particular Lot 6 that is shown in Figure 5.5 is crossed by a navigable river. The northernmost tier of lots is numbered 1 through 4 from east to west. Then the second tier of lots is numbered west to east starting with Lot 5, which is directly south of Lot 4. East of Lot 5 is an aliquot part, a quarter-quarter. East of that part is the fractional Lot 6. It cannot be an aliquot part because a corner of it is inside the meander line that follows the northern bank of the river. East of Lot 6 is Lot 7 and it is also partially in the river. The east boundary of Lot 7 is the section line, so the numbering of the fractional lots drops south the next tier and proceeds east to west again. South of Lot 7 the area is in the river, but directly west of the river is Lot 8. West of Lot 8 is an aliquot part followed by Lot 9 on the western boundary of the section. The numbering scheme drops south to the next tier and Lot 10. It then proceeds from west to east, jumps across the aliquot part, and then Lot 11 is numbered. The last lot numbered is Lot 12 in the southeast corner of the section. It is interesting to note that Lot 12 includes a small area that could be in the tier of lots to its north. However, one of the rules of the creation of fractional lots specifies that very small lots will not be protracted, but such an area will instead be included into adjoining lots, as is illustrated here.

Naming Aliquot Parts and Corners

As mentioned, aliquot parts of a section are not numbered. They are named. Some aspects of their names are intuitive. For example, a 160-acre portion of a section north and east of its center can simply be called the northeast quarter. It can also be abbreviated as NE¼. Describing any of the other quarters of a section can be done similarly, such as NW¼, SW¼, and SE¼. Describing quarter-quarters of a section is done with a combination of abbreviations. For example, the 40-acre portion of the NW¼ that is in the most northwesterly corner of the section is known as NW¼ NW¼, as seen in Figure 5.6. This description can be read as the northwest quarter of the northwest quarter. The other quarter-quarters in the NW¼ could be described with the same system as NE¼ NW¼, SE¼ NW¼, and SW¼ NW¼. There are a couple of

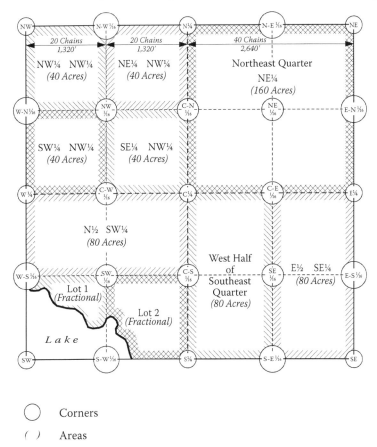

FIGURE 5.6
A section—640 acres.

important keys to such an abbreviation system. First, the smaller tract is always first, meaning it is on the left. The larger tract is on the right. Second, there is no comma between the two elements. A comma would mean "and" in this context. In other words, NE¼, NW¼ would mean the northeast quarter *and* the northwest quarter, whereas the intention of NE¼ NW¼ is to indicate the northeast quarter *of* the northwest quarter. The general structure of the system is that the term on the left is a portion of the term on the right. The same ideas can be used to describe the east half of the southeast quarter. It could be described as E½ SE¼. Fractional lots can be described using their numbers: Lot 1, Lot 2.

The abbreviated descriptions of lots and aliquot parts do not fully define them, however. Writing a complete and correct description requires the addition of the number of the section, the township, the range, and the principal meridian. For example, the S½ SE¼ Section 36, Township 3 South, Range 68 West of the 6th Principal Meridian is complete, concise, and unambiguous. It can only refer to one parcel of land.

The naming of section corners in PLSS is also straightforward. The northeast, northwest, southeast, and southwest corners of a section are simply abbreviated as NE, NW, SE, and SW, respectively. The quarter corners too are simply stated. The corners at the midpoints on the section lines are the north quarter corner, east quarter corner, south quarter corner, and west quarter corner, which are abbreviated as N¼, E¼, S¼, and W¼, respectively. The center quarter, which is at the intersection of the quarter-section lines, is abbreviated as C¼.

It is important to mention that another common name for a quarter-quarter of a section is a sixteenth. Therefore, corners of quarter-quarters are often called sixteenth corners, such as the corners in the middle of quarter sections, where the quarter-quarter lines intersect.

For example, as shown in Figure 5.6, the corner in the centers of the NE¼ of the section common to the four quarter-quarters is called the NE^1/$_{16}$. The other ^{1}/$_{16}$ corners in the centers of the quarter sections are also named for the quarter in which they are set. They are called NW^1/$_{16}$, SE^1/$_{16}$, and SW^1/$_{16}$. The abbreviations for the ^{1}/$_{16}$ corners on the quarter lines of the section refer first to the fact that the lines on which they fall are the centerlines of the section. For example, the corner on the N–S centerline between the NE¼ and the NW¼ is called C-N^1/$_{16}$. The other center quarters are abbreviated in the same pattern: C-E^1/$_{16}$, C-S^1/$_{16}$, and C-W^1/$_{16}$. Finally, the ^{1}/$_{16}$ corners on the section lines themselves follow a similar scheme. Their abbreviations refer to the section line on which they stand first, and then whether they are east or west from the quarter corner on the same section line. Therefore, on the north line of the section, there are the N-E^1/$_{16}$ and N-W^1/$_{16}$ corners. On the east section line, there are two ^{1}/$_{16}$ corners: E-N^1/$_{16}$ and E-S^1/$_{16}$. The abbreviations applied to the ^{1}/$_{16}$ corners on the south section line and the west section line are created with similar logic.

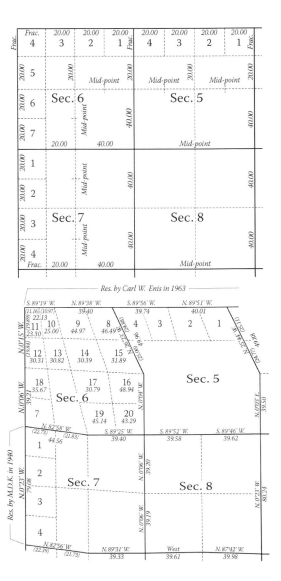

FIGURE 5.7
The plan and the actual.

The upper portion of Figure 5.7 shows the normal subdivision of sections 5, 6, 7, and 8 per the design of PLSS, and the lower portion shows a portion of a township in Wyoming, T18N R89W of the Sixth Principal Meridian, as it actually stands. Portions of the west boundary, north boundary, south boundary, and the subdivisional lines were the subject of a dependent resurvey by a supervisory cadastral surveyor in 1963 under special instructions that restored some lost corners to their true original locations. The work was concerned with protecting the property rights that had arisen from the original survey as well. For example, the fractional lots are shown as given on the original plat of the township, which was approved in 1884. In other words, the surveyor in 1963 did everything he could to reconstruct the positions of the lines and corners as they had existed after the original survey of 1884. And this is consistent with the mandate for local surveyors throughout the states that were carved from the public domain. When working in the PLSS, the objective is always to preserve or reestablish the original locations of the lines and corners if at all possible.

However, by comparing the two parts of Figure 5.7, one can see that there is often a difference between the pure design of the PLSS and the actual section lines and corners as the original surveyor laid them out on the ground. Nevertheless it is certain that the monuments set and lines run on the ground by that original surveyor are conclusive. The corner monuments originally set and the section lines originally run are inviolable. Here are some of the fundamental rules that control in the PLSS system and underline the point:

> First: That the boundaries of the public lands, when approved and accepted, are unchangeable.
>
> Second: That the original township, section and quarter-section corners must stand as the true corners which they were intended to represent.
>
> Third: That the quarter-quarter-section corners not established in the original survey shall be placed on the line connecting the section and quarter-section corners, and midway between them, except on the last half mile of section lines closing on the north and west boundaries of the township, or on the lines between fractional or irregular sections.
>
> Fourth: That the center lines of a section are to be straight, running from the quarter-section corner on one boundary to the corresponding corner on the opposite boundary.
>
> Fifth: That in a fractional section where no opposite corresponding quarter-section corner has been or can be established, the center of such section must be run from the proper quarter-section corner as nearly in a cardinal direction to the meander line, reservation or other boundary of such fractional section, as due parallelism will permit.
>
> Sixth: That lost or obliterated corners of the approved surveys must be restored to their original locations whenever this is possible. (BLM 1979, 6)

It is interesting to note the practical effect of these rules. Section lines and quarter-section lines are correct as they were originally run, even though typical regular section lines are not 1-mile long, and quarter-section lines are not usually a half-mile long. It is usual that the quarter corners of a section are not actually midway between section corners, nor are they in line with section corners. Nevertheless, these monuments are the corners they are intended to represent, exactly. Despite the fact that section lines most often do not have a cardinal bearing and the legal center of the section is not the geometric center of the section, the lines as originally surveyed are right. Even though none of the angles at the section corners is 90° and the opposite sides of the section are not parallel to each other, the lines as originally surveyed are conclusive. Few sections actually contain 640 acres; in fact, other legal subdivisions rarely contain their theoretical acreages. And the actual dimensions of a section measured today seldom match the original record dimensions. But despite all these deviations from the plan, the section is absolutely correct as it was originally laid out on the ground. This overarching principle was codified long ago, by the Act of February 11, 1805. Corners and lines established during the original government survey are unchangeable.

That is not to say that there are no limits to how far a section, township, or quadrangle may depart from the ideal in the PLSS. The general rectangular limits are:

1. For alinement, the section's boundaries must not exceed 21' from cardinal in any part, nor many opposite (regular) boundaries of a section vary more than 21'.
2. For measurement, the distance between regular corners is to be normal according to the plan of survey, with certain allowable adjustments not to exceed 25 links in 40 chains. Township exteriors, or portions of exteriors, are considered defective when they do not qualify within the above limits. It is also necessary, in order to subdivide a township regularly, to set a third limit, as follows:
3. For position, the corresponding section corners upon the opposite boundaries of the township are to be so located that they may be connected by true lines which will not deviate more than 21' from cardinal. (BLM 1973, 70)

However, it should be noted that these rules apply to work done under the current *Manual of Surveying Instructions 1973*. If the surveying were done under other older instructions—and most of the public domain was subdivided long ago—then the work is not necessarily bound by these limits. Along that line, it is important to remember that the rules that governed the original establishment of the corners and lines in the PLSS have not always been exactly as they are today. For example, there was no accommodation of the convergence of meridians, and sections were numbered sequentially from south to north in the original seven ranges of sections laid out in 1785 in Ohio. Nevertheless, no attempt would be made to alter those sections to harmonize with rules in place today. For example, prior to 1855, the distance between standard parallels was often 30 miles or more. In some cases, the parallels were as much as 60 miles apart. When working with townships and sections established in that era, it is important to recognize that they are correct, even though the current instructions place the distance between standard parallels at 24 miles. There are many differences between current guidelines and those of the past.

Here is an example of the evolution of section and quarter-section corners. It is only one instance of official PLSS surveying that was done contrary to the current plan, but was nevertheless absolutely correct at the time. In the Act of May 18, 1796, there was an economy measure. Only every other township was actually subdivided into sections. And in those townships that were subdivided, only every other section line was run on the ground. In other words, in half of the townships, the section lines were run every 2 miles, and corners were marked on those lines every mile.

The Act of May 10, 1800, was passed to amend the Act of May 18, 1796. It had proven impractical to expect the purchasers of sections to set the many corner monuments that the government surveyors had not established. They were stubbing them in by the easiest methods, and that was producing large errors. Therefore, in 1800, Congress mandated that section lines would henceforth be run every mile, not every 2 miles. Corner monuments would be set each mile on the north-south lines and every half mile on the east-west lines. Thereby, the system evolved from setting no quarter corners to setting only those on latitudinal lines. Also, the excesses and deficiencies of the measurements would be placed in the last half mile on the north and west boundary of the township. It took another 5 years, and the Act of February 11, 1805, for the quarter corners on the meridional section lines to be called for.

It follows that, when working with sections that were established before 1805, it might be valuable to recall that all the quarter corners were not set in the original survey. This is only one of many discrepancies from the current instructions that guide the PLSS today.

Another example: the Act of June 1, 1796 mandated the establishment of townships that were, and still are, 5 miles on a side instead of 6 miles. Congress had segregated a portion of land known as the U.S. Military Reserve to satisfy the military warrants it had issued to pay Revolutionary War soldiers. These warrants were in multiples of 100 acres, and 6-mile townships contain 23,040 acres, not evenly divisible by 100. The 5-mile townships contained 16,000 acres and were much more convenient to the purpose. And that is not the only region in which special instructions stipulated deviations from the normal plan.

There are two eminent books that are reliable guides in this regard. *The Manual of Surveying Instructions 1973* and *A History of the Rectangular Survey System* are both published by the U.S. Government Printing Office. The former book was compiled under the auspices of the Bureau of Land Management, Department of the Interior, and is a clear, concise manual that sets out the guidelines for the current Public Land Survey System plan. The latter book was written by C. A. White and is also published by the U.S. Government Printing Office. It is a valuable resource that explains the rules and regulations that governed the PLSS from the beginning to today. It also includes reproductions of many of the former manuals and instructions issued by surveyors general in the past. Together, these books provide practical and valuable insights to both the past and the present PLSS.

Finally, the Public Land Surveying System was the most extensive surveying project ever undertaken. It has been extraordinarily successful. It has evolved. The rules governing its execution have been practical. Still, there are large portions of mountains, wilderness areas, and deserts that have not been surveyed in the PLSS. Unfortunately, unofficial township and section lines are sometimes drawn over and across these areas to provide the continuation of the system where it has never actually been applied by government survey. As has been described here, the general PLSS plan does not necessarily correspond exactly to the actual corners and lines on the ground, but it is best to recall that it is always those actual corners and lines, as originally surveyed on the ground, that are correct and conclusive. Where no such corners and lines were ever run, the PLSS system is not in place at all.

References

BLM (Bureau of Land Management). 1973. *Manual of surveying instructions 1973.* Washington DC: U.S. Government Printing Office.

————. 1979. *Restoration of lost or obliterated corners and subdivision of sections.* Washington, DC: U.S. Government Printing Office.

GLO (General Land Office). 1881. *Instructions of the commissioner of the General Land Office to the surveyors general of the United States relative to the survey of the public lands and private land claims, May 3, 1881.* Washington, DC: U.S. Government Printing Office.

————. 1890. *Manual of surveying instructions for the survey of public lands of the United States and private land claims, January 1, 1890.* Washington, DC: U.S. Government Printing Office.

————. 1902. *Manual of surveying instructions for the survey of public lands of the United States and private land claims, January 1, 1902.* Washington, DC: U.S. Government Printing Office.

————. 1930. *Manual of surveying instructions for the survey of public lands of the United States and private land claims, 1930.* Washington, DC: U.S. Government Printing Office.

White, C. Albert. 1983. *A history of the rectangular survey system.* Washington, DC: U.S. Government Printing Office.

Exercises

1. Which of the following general statements about initial points in the PLSS is incorrect?

    a. Most of the initial points were already in place when they were first mentioned in a manual of instruction.

    b. A latitude and longitude were chosen and an initial point set at that coordinate before the PLSS survey was commenced in a region.

    c. The geographic coordinates of the initial points were determined astronomically.

    d. Several of the initial points still retain their original monumentation.

2. Which of the following states has no initial point within its borders?

    a. Wyoming

    b. Oregon

    c. Iowa

    d. Montana

3. Which of the following categories of corner monuments was not and is not set during an official government survey, unless it is done under special instructions?

    a. Section corners

    b. Quarter-section corners

    c. Township corners

    d. Quarter-quarter corners

4. Under normal conditions and by current instructions, which of the following statements is true concerning both the subdivision of a quadrangle into townships and the subdivision of a township into sections?

    a. The survey begins on the south line, and corner monuments are set every 40 chains on parallels of latitude and meridians of longitude.

    b. The official units of the survey are chains and corner monuments as set by the original surveyor are inviolable.

    c. The random and true line procedure is used on the latitudinal lines, and the east and south lines are the governing boundaries.

    d. The survey begins on the south line; corner monuments are set every 40 chains; and the lines are intended to be parallel.

5. Which of the following have remained unchanged since the beginning of the PLSS?

    a. The pattern of numbering sections in a township

    b. The method of subdividing a township into sections

    c. The size of a township

    d. The principle that land in the public domain must be surveyed before it is sold

6. Under normal conditions and by current instructions, which section's boundaries are completed last in the subdivision of a regular township?

    a. Section 1

    b. Section 6

    c. Section 31

    d. Section 36

7. An original corner monument that was approximately 72 miles west and 95 miles north from an initial point is most probably which of the following?

    a. It is a closing corner at the intersection of the Third Guide Meridian West and the Fourth Standard Parallel North.

    b. It is a standard corner on the Third Guide Meridian West at the southwest corner of Section 6, Township 16 North, Range 12 West.

    c. It is a standard corner on the Fourth Guide Meridian West at the southwest corner of Section 6, Township 12 North, Range 15 West.

    d. It is a closing corner at the intersection of the Third Standard Parallel North and the Fourth Guide Meridian West.

8. What aliquot part of a section has the following corners; C¼, NW^1/$_{16}$, C-N^1/$_{16}$, and C-W^1/$_{16}$?

    a. E½ SE¼

    b. NW¼

    c. NW¼ NW¼

    d. SE¼ NW¼

9. Which of the following statements about the difference between a quarter-quarter of a section and a fractional lot is correct?

    a. A quarter-quarter of a section contains 40 acres; a fractional lot does not.

    b. A quarter-quarter of a section is an aliquot part; a fractional lot is not.

    c. A quarter-quarter of a section is numbered; a fractional lot is not.

    d. The corners of a quarter-quarter section are set during the original government survey; the corners of a fractional lot are not.

10. When a standard, township, or section line intersects a meander line along the high-water mark of a navigable body of water, what sort of a corner is established?

    a. A section corner

    b. A closing corner

    c. A meander corner

    d. A standard corner

Explanations and Answers

1. Explanation:

Initial points are the origins of the PLSS. There are 32 in the con-terminous United States, 5 in Alaska, and several others still have their original monuments. When they were first mentioned in the *Manuals of Instructions* in 1881, most of the 37 points were already in place. However, the criteria for setting initial points never included the idea that they needed to be set at a particular latitude and longitude. Their latitudes and longitudes were determined astronomically and recorded, but they were not placed at previously determined coordinates.

Answer: **(b)**

2. Explanation:

PLSS surveys in Iowa are referred to the Fifth Principal Meridian, which is in Arkansas. However, there is no initial point in Iowa. There are several other states in which there is no initial point.

Answer: **(c)**

3. Explanation:

Quarter-quarter corners are not set under current instructions, and they have not been previously mandated. However, special instructions can be issued that would stipulate that such corners be monumented during the official survey.

Answer: **(d)**

4. Explanation:

While it is true that the subdivision of a quadrangle into townships and the subdivision of a township into sections both begin on the south line, only the boundaries of townships are intended to be geographical lines. The section lines in a township are intended to be parallel to the south and east boundaries of the township. Corner monuments are typically set every 40 chains, and chains are the official units of the PLSS. Original monuments and lines established during the original survey are inviolable.

Answer: **(b)**

5. Explanation:

Until 1796, the numbering of sections in a township was done in a north-south scheme; since then, the east-west zigzag pattern used today has been unchanged. Also, in 1796, only every other township was actually subdivided into sections. Moreover, in the half of the townships that were subdivided, section lines were run only every 2 miles, and corners were marked on those lines every mile. Under the Act of June 1, 1796, 5-mile townships were surveyed in the land set aside as the U.S. Military Reserve to satisfy the military warrants Congress had issued to pay Revolutionary War soldiers. However, the principle that land must be surveyed before it is sold has been fairly consistent in the history of the PLSS.

Answer: **(d)**

6. Explanation:

Under normal circumstances, the subdivision of a township into sections begins at the southwestern corner of Section 36 and continues through the eastern range of sections. That work is followed west consecutively by each of the six ranges of sections in the township. Therefore, the last section completed is Section 6 in the northwestern corner of the township.

Answer: **(b)**

7. Explanation:

There are 24 miles between guide meridians. Therefore, the distance from the principal meridian to the Third Guide Meridian West is 72 miles. The corner in the question is on the Third Guide Meridian West. There are also 24 miles between standard parallels. Therefore, the distance from the baseline to the Fourth Standard Parallel North is 96 miles. The corner in the question cannot be on the Fourth Standard Parallel North. However, it is in the township whose northwest corner is at the intersection of the Third Guide Meridian West and the Fourth Standard Parallel North. That is Township 16 North and Range 12 West. It is the corner in that township 1 mile south of the Fourth Standard Parallel North. It is the southwest corner of Section 6.

Answer: **(b)**

8. Explanation:

The 40-acre aliquot part known as the SE¼ of the NW¼ has the corners described.

Answer: **(d)**

9. Explanation:

A quarter-quarter of a section does not necessarily contain 40 acres, and it is possible for a fractional lot to contain 40 acres and still not be an aliquot part. A fractional lot is a subdivision of a section that is not a half or a quarter of the previously larger part of a section. On the other hand, a quarter-quarter of a section is an aliquot part. A fractional lot is numbered; a quarter-quarter of a section is not. Finally, neither the corners of the quarter-quarter of a section nor the corners of a fractional lot are monumented during the original survey unless such work is specified by special instructions.

Answer: **(b)**

10. Explanation:

Meander lines do not describe legal boundaries. They are used to set off certain bodies of water from the PLSS. Meander corners are established where the meander line along the mean high water is intersected by a standard, township, or section line.

Answer: **(c)**

Index

An environmentally friendly book printed and bound in England by www.printondemand-worldwide.com

PEFC Certified

This product is
from sustainably
managed forests
and controlled
sources

www.pefc.org

This book is made entirely of sustainable materials; FSC paper for the cover and PEFC paper for the text pages.

#0112 - 281014 - C0 - 234/156/11 [13] - CB